Danubia Moreno
Wagner Vilano

Zooplankton in the Rift Zone of the Brazilian Southeast Coast

AF666520

Danubia Moreno
Wagner Vilano

Zooplankton in the Rift Zone of the Brazilian Southeast Coast

Temporal and Nictemeral Variation

ScienciaScripts

Imprint

Any brand names and product names mentioned in this book are subject to trademark, brand or patent protection and are trademarks or registered trademarks of their respective holders. The use of brand names, product names, common names, trade names, product descriptions etc. even without a particular marking in this work is in no way to be construed to mean that such names may be regarded as unrestricted in respect of trademark and brand protection legislation and could thus be used by anyone.

Cover image: www.ingimage.com

This book is a translation from the original published under ISBN 978-3-330-76803-1.

Publisher:
Sciencia Scripts
is a trademark of
Dodo Books Indian Ocean Ltd. and OmniScriptum S.R.L publishing group

120 High Road, East Finchley, London, N2 9ED, United Kingdom
Str. Armeneasca 28/1, office 1, Chisinau MD-2012, Republic of Moldova, Europe
Managing Directors: Ieva Konstantinova, Victoria Ursu
info@omniscriptum.com

Printed at: see last page
ISBN: 978-620-8-58329-3

Copyright © Danubia Moreno, Wagner Vilano
Copyright © 2025 Dodo Books Indian Ocean Ltd. and OmniScriptum S.R.L publishing group

CONTENTS

I dedicate this work to Wagner F. Vilano, my

companion in life, in this and so many others, my

soul partner, my best friend. To whom

strength, tranquility and hope.

The one who, in the midst of the storm, is able to keep my boat stable.

the stability of my boat. The one who makes me create

past, enjoy the present and inspire the future.

I LOVE YOU

ACKNOWLEDGMENTS

Finally, this journey has come to an end thanks to the encouragement, goodwill and advice of special people who have come into my life over the last few years.

I would first and foremost like to thank Wagner Ferreira Vilano who, although unofficially, was more than just the co-supervisor of this work, where he applied some of the knowledge he had acquired throughout his life with patience and total dedication. Through his endless guidance, he enabled me to gain more wisdom and practice, broadening my view of science and my field of professional activity. The completion of this stage in my life was only possible because of his support and enthusiasm from start to finish.

To Dr. Teodoro Vaske Junior, for the trust he placed in me when, without even knowing me, he agreed to guide me and provide his research material. His attention, calmness and wisdom were essential at various times during this journey.

To the students of Prof. Dr. Teodoro, who helped several times during the field work, sparing their efforts in the collections and thus being fundamental for this work to be possible.

To the members of the qualifying and master's examining boards, Dr. Marcelo Antônio Amaro Pinheiro and Dr. Acàcio Ribeiro Gomes Tomàs, for their criticism and suggestions.

To Dr. Patricio A. Hernàez Bové and Ms. Renata Rodrigues Melo (Topanga) for their attention and willingness to help with the statistical analysis.

Finally, I would like to thank my family for their advice, teachings and sacrifices over the years: my mother Sônia, my father Darcy Moreno, my second mother and friend Tereza Valério, my brothers Davinton and Débora Moreno.

To Unesp IB/CLP for welcoming us and making the laboratories and equipment available, to the staff of the Didactic Laboratory, Marcia, Luciana, Aline, Willian who always made themselves available when we used the laboratories.

Summary

Shallow coastal areas, such as sandy beach breakwaters, are little studied when compared to other coastal ecosystems. However, they are considered to be of great ecological importance, as they are home to a high diversity of organisms that feed, reproduce and seek protection from predators. The lack of studies is mainly due to the high level of hydrodynamics, which makes it difficult to sample, especially at night. This study analyzed the temporal and nictemeral variation of the zooplankton fauna in the surf zone of Praia da Aviaçao, a densely inhabited beach located in the municipality of Praia Grande on the central coast of the state of São Paulo. The organisms were collected with a plankton net by dragging horizontally, superficially and perpendicular to the beach at a depth of 1m. Temperature and salinity data were also obtained along with visual observations of the state of sea agitation, local wind and cloudiness. This process took place bi-monthly between June 2013 and April 2015 during the day and night periods in order to assess the variation in zooplankton composition and its relationship with vertical migrations. The zooplankton of the beach analyzed was represented by 116 taxa, including holoplankton, meroplankton, benthic organisms (present in the water column due to high turbulence) and bentoplankton (organisms that migrate vertically). The composition of the fauna was very diverse and, throughout the study, there was little variation in diversity between sampling periods, apart from a slight seasonal variation. The total abundance of planktonic organisms caught during the day ranged from 384 to 14746 org.m-^{3} (4153.1± 5766 org.m-^{3}) and at night from 504 to 67269 org.m-3 (7845.3± 18844 org.m-3). The zooplankton community showed a significant difference ($p < 0.001$) in terms of abundance, with the holoplankton fauna being more abundant than the meroplankton fauna and both in relation to the bentoplankton fauna. The greatest abundance and percentage of contribution among the holoplankton was attributed to the Copepoda, Cladocera and Tunicata groups. As for the meroplankton, the Mollusca and Echinodermata groups stood out and, among the bentoplankton, the Mysida group. The zooplankton fauna showed a temporal variation in abundance, with significant differences ($p < 0.001$) in each month of sampling, with the values recorded in summer being similar to each other (85%), but different from the other seasons (< 40%). However, there were no significant differences in abundance (p = 0.299) related to the environmental variable water temperature. A greater number of organisms were recorded in the water column during the night, with an increase in fauna, mainly due to the greater frequency of misidaceans, decapod larvae and juvenile fish. However, the greater occurrence of these organisms at night did not statistically reflect differences in the total abundance of fauna recorded during the day (p = 0.123). This study confirmed that the upwelling zone of Praia da Aviaçao, in Praia Grande (SP), is composed of an abundant, rich and extremely dynamic zooplankton community, reflecting the joint action of abiotic and biotic processes responsible for its maintenance in this coastal environment.

Key words: Vertical migration, nictemeral variation, surf zone, zooplankton.

1. introduction

Despite occupying a narrow portion of the planet's surface, sandy ocean beaches cover the largest part of the world's coastal region and are of major socio-economic and environmental importance (Brown & McLachlan, 1990). According to Ab'Saber (2001), Brazil has the largest group of sandy beaches in the world and, along the entire coastline, sandy beaches are considered to be one of the most extensive environments, occupying a stretch of 6000 km, with 430 km corresponding to the São Paulo coast (Souza & Suguio, 1996).

Resulting from the interaction of geological, geomorphological, climatic and oceanographic factors, the beach environment plays an important role in protecting the coastline, as it adjusts to hydrodynamic conditions, due to its composition of unconsolidated material that is highly mobile, acting as buffer zones that protect the continent from the direct action of ocean energy (Muehe, 1994; Souza, 1997; Hoefel, 1998). This ecosystem is also important because it is home to a considerable diversity of organisms (Brown & McLachlan, 1990), which are often used by humans as a source of food and income.

Due to constant environmental changes, the beach ecosystem is considered to be highly dynamic and variable over time and space (Short, 1983) and, depending on the morphological pattern and local hydrodynamic regime, beaches can be classified according to their morphodynamic pattern, ranging from dissipative to reflective (Wright & Short, 1984). According to Hoefel (1998), beaches can also be delimited into sub-environments due to the hydrodynamic processes at work, which are capable of modifying the morphological configuration of each beach section. The surf zone stands out among the other existing compartments, as it is considered to be of great importance due to the high energy and turbulence in this region of the beach profile (McLachlan, 1980).

Delimited between the outer limit of the wave break and the beach line (McLachlan *et al.,* 1981; Carter, 1988; Brown & McLachlan, 1990), the surf zone receives the direct action of the waves generated in the ocean, which is one of its important functions, since it is where the incident waves break and their energy is dissipated on the beach (Suguio, 1992; Souza, 1997; Hoefel, 1998). This hydrodynamics promotes sediment transport, both towards the sea and towards the coast, favoring changes in beach morphology and the development of the coastline (McLachlan, 1980; Suguio, 1992; Calliari *et al.*, 2003; Davidson-Arnott, 2010). The action of waves and coastal currents, in addition to transporting sediments within the upwelling zone, is also capable of promoting the concentration of debris, phytoplankton and the recirculation of organic matter adhered to the unconsolidated sandy material (Brown & Mclachlan, 1990; Rorig *et al.*, 1997; McLachlan & Brown, 2006; Ferreira, 2007). This situation is favored as a result of the high concentration of nutrients present in the upwelling zone due to its boundary location between the terrestrial and marine environments, which promotes exchange between the two systems (Buthelezi, 2012).

Thus, the hydrodynamics of the area are considered to be the driving force behind biological processes, as they are able to make organic matter available in the water column, allowing for greater bacterial decomposition and the reuse of nutrients by phytoplankton (McLachlan, 1980; Rorig *et al.*, 1997). This generates a chain cycle that promotes the development of zooplankton organisms and, consequently, fauna in general (McLachlan *et al.*, 1981; McLachlan & Erasmus, 1983; McLachlan & Brown, 2006).

In this sense, understanding the zooplankton fauna is extremely important for understanding the structure of the coastal marine ecosystem (Lopes *et al.*, 2006), as they play an important role in the marine pelagic food chain, as prey, competitors and consumers, transferring energy and carbon between primary producers and the upper trophic levels, as well as cycling nutrients in the water column (Lasiak, 1986; Brown & Mclachlan, 1990; Brandini *et al.*, 1997; Lenz, 2005; Lopes *et al.*, 2006). Specifically in the surf zone, they are considered the base of the food chain for a diversity of juvenile fish and crustaceans (Lasiak, 1986; Brown & McLachlan, 1990; Beyst *et al.*, 2001a, 2001b; 2002a), many of which are used commercially (Paiva-Filho & Toscano, 1987; Bertozzi, 2002; Gaelzer & Zalmon, 2003; Silva *et al.,* 2004).

However, the dynamics and high turbulence of the surf zone make it a highly complex environment (Lasiak, 1986), requiring various adaptations on the part of the zooplankton and communities in general (Wooldridge, 1983; Brown & McLachlan, 1990). However, the region is capable of harboring a varied fauna of fish and marine invertebrates that use the area as a habitat, feeding and recruitment points, or even as protection against predators (McLachlan *et al.*, 1981; Lasiak, 1981, 1986; Clark *et al*, 1996b; Hams, 1996; Beyst *et al.*, 1999, 2002a; Pessanha *et al.*, 2000; Gaelzer & Zalmon, 2003; Pessanha & Araujo, 2003; Barreiros *et al.,* 2004; Silva *et al.*, 2004; McLachlan & Brown, 2006; Vilano, 2013; Favero & Dias, 2013, 2015).

The zooplankton fauna of the surf zone on sandy beaches is made up almost entirely of crustaceans, although other organisms can occur sporadically (Bersano, 1994; McLachlan & Brown, 2006; Buthelezi, 2012). The zooplankton in this region can be classified as resident, regularly present species, and non-resident, composed of species that occur migratory or sporadically (Clark *et al.*, 1996a; McLachlan & Brown, 2006). In both classifications, organisms can occur in the form of eggs, larvae, juveniles and adults (Modde, 1980; Mees & Jones, 1997; Beyst *et al.,* 1999, 2001a, 2002a, 2002b; Barreiros *et al,* 2004; Dominguez Granda *et al.,* 2004; Silva *et al.*, 2004; McLachlan & Brown, 2006; Crawley *et al.*, 2006; Vasconcellos *et al.*, 2007; Avila *et al.*, 2009; Sahu *et al.*, 2012; Favero & Dias, 2013).

Despite being considered relatively homogeneous, studies show that there can be major changes in the composition and richness of the zooplankton fauna in the surf zone, directly related to changes in physical and chemical variables (temperature, salinity, sediment granulometry, currents, tide, wave height, turbidity and beach slope) and temporal patterns (seasonality, time of day and moon phase) (Romer, 1990; Clark *et al,* 1996a; Clark, 1997; Dye *et al,* 1981; Dexter, 1984; Hamerlynck & Mees, 1991; Mees & Jones, 1997; Lock *et al,* 1999; Beyst *et al*, 2001b, 2002a; Watt-Pringle & Strydom, 2003; Dominguez Granda *et al.,* 2004; Miller, 2004; McLachlan & Brown, 2006; Crawley *et al.*, 2006; Vasconcellos *et al.*, 2007; Favero & Dias, 2013, 2015). However, variations in the composition of zooplankton species can also occur as a result of the pattern of daily vertical and horizontal migration associated with spawning periods, diet and predation (Moran, 1972; Rossouw, 1983; Webb & Wooldridge, 1990; Warman *et al.,* 1991; McLachlan & Brown, 2006).

On the Brazilian coast, although sandy beaches are considered to be one of the most extensive environments, studies that seek to understand the behavior of the zooplankton community in subtidal zones are still very scarce, unlike other regions of the world such as South Africa, Belgium and the United States (McLachlan & Brown, 2006). In Brazil, among the studies describing the fauna found in coastal environments, few specifically analyze the upwelling zone, and most are restricted to the ecology of a single species or taxonomic group.

Therefore, taking into account the gaps in existing knowledge and the importance of zooplankton in shallow areas of high turbulence, this study aims to document the zooplankton community of an exposed sandy beach located on the central coast of São Paulo.

2. Objectives

2.1. General Objective

To describe the zooplankton community present in the surf zone of an exposed sandy beach, located in the city of Praia Grande, Metropolitan Region of Baixada Santista (SP).

2.2. Specific Objectives

a. To determine the zooplankton composition in the surf zone of Praia da Aviaçao, Praia Grande (SP);

b. To analyze the structure of the zooplankton community and its qualitative and quantitative variation on a temporal (over two years) and nictemeral (day and night) scale;

c. Determine the total zooplankton abundance and that of the main taxonomic groups over time and day;

d. Characterize the main taxonomic groups in terms of frequency of occurrence and percentage of contribution;

e. Evidence of the ecological structure of zooplankton through diversity and evenness indices.

3. Background

3.1. Sandy beaches

The nomenclatures used to define the beach ecosystem differ considerably. However, the most notable characteristic is the presence of non-cohesive sediments that beaches constitute, as well as the action of hydrodynamic factors acting on this system (Hoefel, 1998). According to Suguio (1992), beaches are generally defined as a region made up of unconsolidated material, usually sandy, which develops along water bodies such as oceans, seas or lakes. According to the same author, this ecosystem is limited from the level of mean low tide (the depth where waves interact with the substrate) to the limit of storm waves, or where there are changes in the physiography. This ecosystem is highly dynamic, obeying a constructive and destructive cycle that results from the interaction between oceanographic, climatic and geological factors (Souza, 1997).

Considered extremely fragile, the beach environment is of great ecological, biological and socio-economic importance, due to the existence of a high concentration of biomass (depending on local hydrodynamics) and the presence of countless organisms that make up the marine trophic chain (Brown & McLachlan, 1990), many of which are used by man for consumption or as a source of income (Paiva-Filho & Toscano, 1987; Bertozzi, 2002). Beaches also stand out for their function as a leisure and recreational area, as well as being considered as a protection zone for the coastline against the destructive action of waves (Muehe, 1994; Souza, 1997; Hoefel, 1998).

Generally speaking, the beach is made up of different compartments and, in order to designate each section, various terminologies have been used and discussed over time (Suguio, 1992; Muehe, 1994; Menezes, 1994; Angulo, 1996). According to Tessler & Mahiques (2000), who carried out a review on the use of terms, *there is* only agreement on terminology in relation to the emersed portions of beaches, making it difficult to establish a universal nomenclature to designate the various compartments of the beach strip.

This study adopted the morphological zonation terminology presented by Hoefel

(1998), where the aerial portion of the beach is subdivided into post-beach and beach face, and the submerged part into mid-beach and foreshore (Figure 1). According to Hoefel (1998), the post-beach corresponds to the upper zone of the beach located above the maximum spreading limit until the onset of abrupt physiographic changes. The beach face, which usually refers to the steepest part, is limited to the portion where the waves spread, ebb and flow. The middle beach is located below the mean sea level, where wave breaking occurs. And finally, the foreshore refers to the portion dominated by wave break-up processes extending out into the sea.

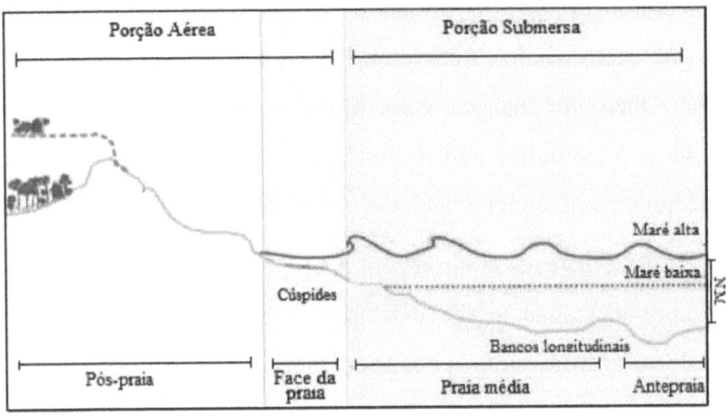

Figure 1: Morphological zone of a sandy beach, according to the classification proposed by Hoefel (1998). Source: Modified from Short (1999) *apud* Moura (2012).

The submerged area of the beach profile can also be defined and subdivided according to the hydrodynamic processes generated by the waves, where different terminologies are used in the literature. This study will adopt the terminology based on the different stages of wave transformation proposed by Aagaard & Masselink (*in* Short, 1999), where the submerged pool is divided into four hydrodynamic regions: the spreading zone, where the waves collide with the beach face; the breaking zone, where there is high turbulence due to the breaking of the waves and dissipation of their energy, this being subdivided into the inner breaking zone where most of the waves break and the outer breaking zone where the waves of greater height break; and finally, the blistering zone of the waves, associated with the reduction in depth (Figure 2).

Beaches are classified according to certain characteristics, such as the level of exposure to waves (protected or exposed), the way in which wave energy is released, and the level of slope of the beach (McLachlan, 1980; Wright & Short, 1983; McLachlan & Dorvlo, 2005). According to these characteristics, beaches have different morphodynamic stages, which can vary from dissipative to reflective, passing through intermediate stages, depending on the morphological pattern and local hydrodynamic regime (Wright & Short, 1984; Hoefel, 1998). The easiest differences to identify between the different morphodynamic stages are related to the breaking zones, where visually one can see extensive zones with several wave breaks, narrow zones with only one break line, or even the absence of one (Mclachlan & Erasmus, 1983; Brown & Mclanhlan; 1990).

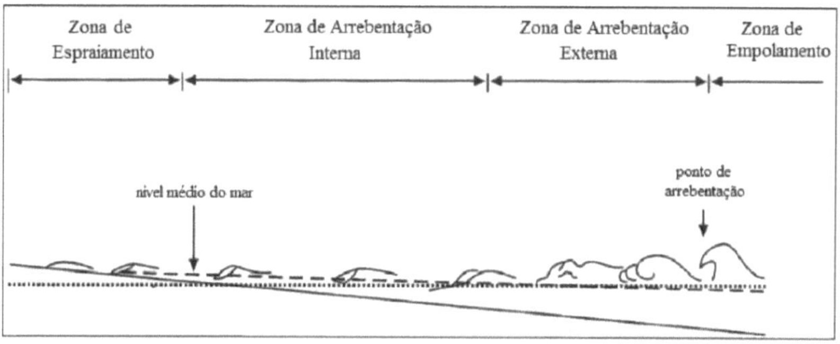

Figure 2: Hydrodynamic zone, according to the classification proposed by Aagaard & Masselink (1999). Source: Modified from Svendsen *et al.*(1978) (*in* Short, 1999).

On the coastline of the State of São Paulo, it is possible to observe beaches that have different morphodynamic stages, where this differentiation is related to the geomorphological features of the coast that are distinguished along the São Paulo coast (Souza, 1997; 2012). Suguio (1992) generally classified beaches as: pocket beaches, dissipative beaches, reflective beaches and transitional beaches. However, due to the different models proposed for the evolution of the beach morphodynamic state, such as the models for the American and Australian coasts (Wright & Short, 1984), Souza & Suguio (1996; 2003) and Souza (1997; 2001) established a specific morphodynamic compartmentalization for the São Paulo coast, such as: dissipative (low or high energy); reflective (low or high energy); intermediate; ultra-dissipative; and mixed (low energy reflective/dissipative).

The transitional stretch between the main geomorphological differences on the São Paulo coast occurs in the Baixada Santista region, which makes the local plateaus highly complex. Unlike the northern part of the coast, the plateaus of the Central-Southern region have submerged characteristics, forming continuous, less jagged beaches with a high degree of exposure to waves (Fulfaro & Ponçano, 1974; Ponçano, 1976; Souza, 1997), creating conditions for the establishment of the dissipative morphodynamic pattern.

On beaches with these characteristics, where the dissipative stage predominates, the breaking zone is wide and has a low slope, thus allowing the occurrence of long-period, high-amplitude and high-energy waves that break far from the beach face, decreasing in height as their energy dissipates along the breaking zone (Mclachlan, 1990; Lippmann & Holman, 1990; Souza, 1997; Hoefel, 1998; Calliari *et al.*, 2003). Wave breaking is of the sliding-diving type (Souza, 1997; Hoefel, 1998; Calliari *et al.*, 2003) and there is a predominance of waves considered to be stationary (Wright *et al.*, 1982), i.e. waves that can result from the meeting of two wave trains going in opposite directions causing vertical oscillation of the water surface (Suguio, 1992). Because the sand stock on dissipative beaches is present in the submerged region, longitudinal banks parallel to the beach are quite common and, depending on the hydrodynamic conditions, can migrate along the breaking zone (Wright *et al.*, 1982; Hoefel, 1998; Calliari *et al.*, 2003).

The incidence of waves oblique to the coastline, as well as the action of local winds, often produce currents parallel to the beach, known as longitudinal currents or littoral drift (Wright *et al.*, 1982; Komar, 1991; Suguio, 1992; Souza, 1997; Calliari *et al.*, 2003). The convergence of two longitudinal currents, which flow in opposite directions, can also give rise to return currents that flow from the coast towards the sea (Davis & Hayes, 1984; Suguio, 1992; Souza, 1997; Hoefel, 1998). Both currents are responsible for transporting suspended material, which can be either inorganic (sediment) (Wright *et al.*, 1982; Suguio, 1992; Souza, 1997; Hoefel, 1998) or organic (plankton) (Campbell & Bate, 1997; Bonecker *et al.*, 2002; Ferreira 2007).

3.2. Zooplankton

The marine planktonic community, according to Brandini *et al.* (1997), is made up of heterogeneous taxonomic groups that encompass a wide variety of species, both animal and plant. Within this community, zooplankton is represented by the majority of marine animal phyla, from porifers to chordates (Nibakken, 1993; Lopes, 2007) and, for this reason, corresponds to a highly complex community (Raymont, 1983).

One of the main characteristics of zooplankton organisms is their low swimming capacity. For this reason, they are passively transported by tides and marine currents, and their large-scale distribution is related to the displacement of water masses (Lopes *et al.*, 2006). However, according to Suthers & Rissik (2009), many zooplankton species are able to swim and control their buoyancy and body position, thus reaching different currents and depths.

For Lopes *et al.* (2006), understanding the zooplankton fauna is extremely important for understanding the structure of the marine ecosystem. According to the author, this is due to the important position of this community in the pelagic food web. Among their many functions, these organisms act as prey, competitors and consumers, transferring energy and carbon between primary producers and higher trophic levels, as well as recycling nutrients in the water column (Lasiak, 1986; Brown & Mclachlan, 1990; Brandini *et al.*, 1997; Kurtz, 1999; Lenz, 2005; Lopes *et al.*, 2006; Suthers & Rissik, 2009). In addition to these functions, zooplankton are considered the base of the food chain for a diversity of juvenile fish and crustaceans (Lasiak, 1986; Brown & McLachlan 1990; Beyst *et al.*, 2001a; 2002a), and therefore the recruitment of many species depends on the stock and diversity of the zooplankton community (Kurtz, 1999; Lopes *et al.*, 2006).

Among the different organisms belonging to this community, there are species classified and called holoplankton (euplankton or permanent plankton), i.e. those that complete their entire life cycle in the plankton, and meroplankton (hemiplankton or temporary plankton), corresponding to taxa that remain in the plankton only during the initial stages of their life cycle (eggs and larvae), as when they are adults they make up the benthic or nectonic fauna (Brandini *et al.,* 1997; McLachlan & Brown, 2006; Lopes, 2007; Suthers & Rissik, 2009).

The vast majority of holoplanktonic organisms are crustaceans, mainly copepods (Parsons *et al.*, 1984; Lopes *et al.*, 2006; Suthers & Rissik, 2009), but euphausiids and cladocerans are also common. However, the holoplankton fauna is also made up of other groups such as filter-feeding urochordates, appendiculates and salps, as well as predators such as hydromedusae and ketognaths (Levinton, 1982; Nibakken, 1993; Brandini *et al.,* 1997; Lopes *et al.*, 2006). Meroplankton comprises mainly crustacean larvae, molluscs, polychaetes and echinoderms, as well as fish eggs and larvae (ichthyoplankton) (Brandini *et al.,* 1997; McLachlan & Brown, 2006; Suthers & Rissik, 2009), and according to Lopes (2006; 2007) other groups such as heterotrophic protists, for example foraminifera, also make up marine plankton.

According to Suthers & Rissik (2009), the planktonic fauna can also be classified according to the variation in size of the organisms. Among the different classifications, the one proposed by Sieburth *et al.* (1978) is still widely used today, with the size categories established as micro (20 - 200μ m), meso (0.2-20mm), macro (2-20cm) and megazooplàncton (>20cm).

3.2.1. Zooplankton in the surf zone

The zooplankton fauna that specifically inhabits the surf zone is also composed mainly of crustaceans, but the region is capable of harboring a varied fauna of fish and benthic and planktonic marine invertebrates, with gelatinous organisms being less frequent and sporadic (McLachlan & Brown, 2006).

The different taxonomic groups in the surf zone were recognized and characterized by Brown & McLachlan (1990) into two categories, "resident" and "non-resident", according to their pattern of occurrence. According to the authors, the species considered to be resident correspond to organisms that constantly inhabit the surf zone, represented by misidaceans, small crustaceans and animal larvae that inhabit the sandy substrate. The so-called non-resident fauna is represented by meroplankton (eggs and larval stages of non-planktonic animals), as well as organisms that appear in the surf zone occasionally, as they can be transported from nearby environments, such as estuaries, rocky shores and the open ocean, by wind, waves and currents (Brown & McLachlan, 1990). However, based on the frequency of capture, Clarck (1996a) established the classification of non-resident fauna as "migratory", when the occurrence of organisms follows specific patterns and they are present in the reef, voluntarily or involuntarily, at certain times of the year or in the same months of subsequent years, and the so-called "sporadic" when the occurrence does not follow a specific pattern. According to Brown & McLachlan (1990), the presence of non-resident fauna is quite variable and, although its occurrence is not constant, it is of great importance as a source of food for the other components of the trophic chain.

Some authors also consider some small organisms to be part of the plankton, such as isopods, amphipods and misidaceans, which live close to the sandy substrate, but occasionally migrate to the plankton, as they have a certain swimming capacity. They are considered to be residents of the upwelling zone (McLachlan, 1983; McLachlan & Brown; 2006) and are known by different terms, for example, Omori & Ikeda (1984) define them as epibenthic plankton, Wooldridge (1989) defines them as ticoplanktonic, Mees & Jones (1997) use the term hyperbenthic, Dye *et al.* (1981) and McLachlan & Brown (2006) consider them to be bentoplanktonic. In this study, these organisms will be referred to as bentoplankton, as this is the most recent term and is considered the most appropriate due to their way of life.

In general, the zooplankton fauna of tidal flats can be considered relatively homogeneous. However, studies have shown that species richness and composition undergo changes directly related to physical and chemical variables. Among these factors, the most cited are temperature, salinity, sediment granulometry, beach slope, turbidity, tide and hydrodynamic variations such as currents and wave heights (Dye *et al.,* 1981; Clark *et al.*, 1996b; Clark, 1997; Romer, 1990; Beyst *et al.*, 2001b; 2002a; Watt-Pringle & Strydom, 2003; Dominguez Granda *et al.*, 2004; McLachlan & Brown, 2006; Crawley *et al.*, 2006; Vasconcellos *et al.*, 2007; Favero & Dias, 2015). However, variations in zooplankton fauna can also occur according to temporal patterns, such as seasonality, time of day and lunar phase (Watkin, 1941; Colman & Segrove, 1955; Dexter, 1984; Wooldridge, 1989; Webb & Wooldridge, 1990; Hamerlynck & Mees, 1991; Clark *et al*, 1996a; Mees & Jones, 1997; Lock *et al.,* 1999; Beyst *et al*., 2001a; 2001b; 2002b; Dominguez Granda *et al.,* 2004; McLachlan & Brown, 2006; Favero & Dias, 2013). However, there is still confusion about the real factors that are related to temporal variations, as there are a range of possibilities involving physical and biological variables together (Clark *et al*., 1996a).

With regard to zooplankton biomass, although there are few studies that attempt to estimate this parameter, according to Brown & Mclachlan (1990), in many locations the upwelling zones have higher values than deep waters. In addition, the few studies that have been published report that local biomass can also vary greatly depending on the time of day (day and night) (Mclachlan & Brown, 2006).

In general, the high biomass values are justified by the frequent use of the upwelling zone as a feeding area by zooplankton due to the constant occurrence of phytoplankton blooms, as well as the high amount of suspended detritus available in the region (Clutter, 1967; McLachlan *et al*., 1981; Lewin & Schaefer, 1983; Romer, 1986; Brown & McLachlan, 1990). However, other food sources are usually abundant in these turbulent areas, thus allowing fauna to easily change their feeding pattern when necessary, which is a typical opportunistic characteristic of many organisms that inhabit this region (McLachlan, 1983; Wooldridge, 1983).

In general, the richness and abundance of zooplankton fauna in tidal flats means that this group occupies an important position in the local trophic chain, as it is responsible for attracting primary carnivores and, consequently, juvenile fish, further increasing the local biomass (Vega-Pérez, 1993; Mclachlan & Brown, 2006). However, as well as attracting many organisms due to the high availability of food, the area also provides fauna, such as ichthyoplankton, with shelter from predators that use their vision to feed, as the local characteristics of high turbulence and turbidity make it difficult for predators to act (Mclachlan & Brown, 2006). In this way, the surf zone can be considered a highly important habitat for the development of many species, since it is constantly used as a nursery, providing protection and food for a variety of organisms (Modde, 1980; McLachlan *et al.*, 1981; Lasiak, 1981; 1986; Bennet, 1989; Brown & McLachlan, 1990; Gibson *et al.*, 1993; Monteiro-Neto & Musick 1994; Clark *et al.*, 1996a; Beyst *et al.*, 2001b; Gaelzer & Zalmon, 2003; Pessanha & Araujo, 2003; Silva *et al.,* 2004; McLachlan & Brown, 2006; Vasconcellos *et al.*, 2007; Favero & Dias, 2013; 2015).

However, due to its hydrodynamic characteristics, the surf zone is also considered, according to Lasiak (1986), to be a highly complex environment. This factor requires organisms living in this environment to acquire various adaptations for their survival (Wooldridge, 1983; Brown & McLachlan, 1990). One of these, mentioned above, is opportunistic feeding behavior (Wooldridge, 1983), as well as the characteristic of protecting the young, where some species have marsupials (egg sacs) or the ability to move to quieter areas when spawning, and can also distribute themselves in patches, thus avoiding predation (Cockcroft & McLachlan, 1986; McLachlan & Brown, 2006).

In addition, the high dynamics of the environment require planktonic organisms to be larger and, above all, to have greater swimming capacity when compared to species from less turbulent environments (Clutter, 1967; McLachlan & Brown, 2006). In this way, those with a certain swimming capacity can distribute themselves along the beach according to the food on offer (Wooldridge, 1989) or even the hydrodynamic intensity of the environment, preferring areas closer to or further from the coastline according to this factor (Clutter, 1967; Wooldridge; 1983).

The same swimming capacity allows for the frequent occurrence of vertical and horizontal migrations carried out by many species of crustaceans, especially in their adult phase (McLachlan & Brown, 2006). These migrations are normally carried out by benthoplanktonic species which, like many other organisms, are more active at night (Moran, 1972; Wooldridge, 1981; DeLancey, 1987; Webb & Wooldridge, 1990; Warman *et al.*, 1991). Studies carried out by Rossouw (1983) and Romer (1986) suggest that a large part of migrations are associated with the search for food or the escape from predators, with the variation in fauna being clearly perceptible in diurnal and nocturnal samples (McLachlan & Brown, 2006). Horizontal migration behavior may also be related to spawning patterns, where many species move out of the spawning zone to spawn and when juveniles return to their place of origin in search of food and protection (Cockroft & McLachlan, 1986). However, migrations related to other factors, such as the tidal cycle and the search for more stable areas during storms, may also be frequent, but the difficulties of sampling in adverse conditions of high hydrodynamics in the surf zones prevent effective proof of these events (Cockroft & McLachlan, 1986; McLachlan & Brown, 2006).

According to McLachlan & Brown (2006), in general, the distribution pattern of the zooplankton fauna living in the upwelling zone is usually clear and well defined throughout the sites already studied, but numerous factors may be associated with the spatial distribution of many species (Wooldridge, 1989). The high hydrodynamics associated with the high degree of turbulence in the region is highlighted by Lasiak (1986) as a factor that can affect the distribution and abundance of many species that occupy this environment. For Wooldridge (1989), the occurrence and intensity of return currents and littoral drift can also interfere with the uneven spatial distribution of zooplankton fauna along the coastline, which is directly related to the swimming capacity of the organisms.

3.2.1.1. Zooplankton Studies in Breaching Zones

The zooplankton community of sandy beach breakwaters, especially the macrofauna with a certain swimming capacity (juvenile fish and shrimps), is frequently studied in various regions of the world such as Mexico, Belgium, Holland, Scotland, the United States and, above all, South Africa (McLachlan & Brown, 2006).

In Brazil, the various studies aimed at analyzing the composition and variation of the zooplankton fauna of the coastal region have been carried out in sheltered areas or at depths greater than 5m with the help of boats. However, specific studies to analyze the composition, diversity and variation of the zooplankton fauna of the surf zone in order to generate a better understanding of this environment have been little developed in the country (Figure 3 and Annex 1). For this reason, there are still many gaps in knowledge about the occurrence and distribution of important taxonomic groups belonging to the zooplankton community of coastal areas.

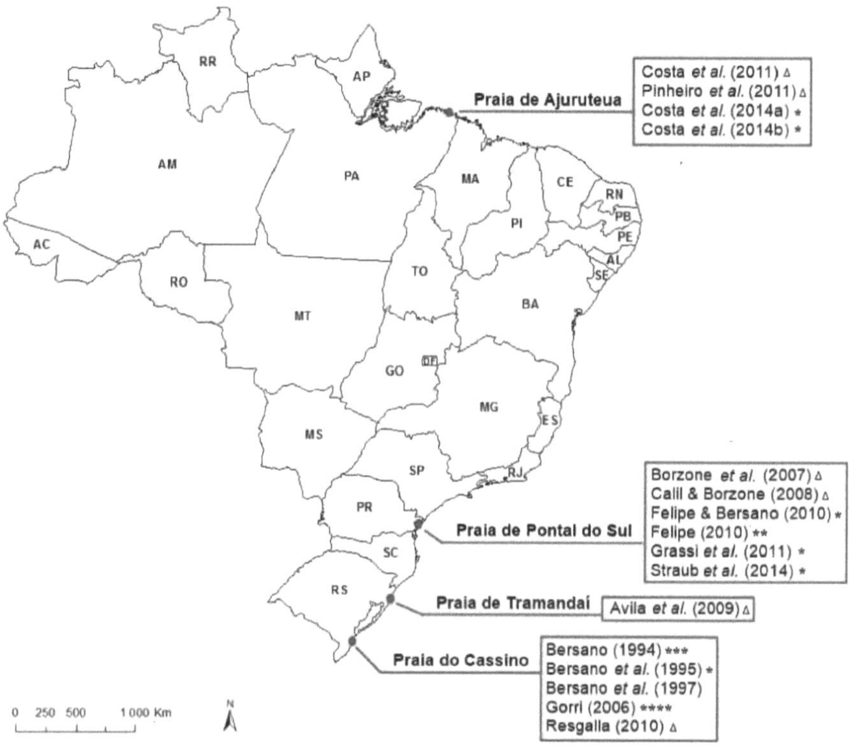

Figure 3: Distribution of studies on the zooplankton of the surf zone carried out along the Brazilian coast (* Congress abstracts; ** Monograph-Graduation; *** Dissertation-Master's degree; **** Thesis-Doctorate; Δ Published articles). Source: Modified from IBGE map base.

The few studies published on the zooplankton of the turbulent zones of Brazilian sandy beaches are still related to a specific taxonomic group or, when they analyze the composition and distribution of the zooplankton community as a whole, they emphasize only a certain taxonomic group. In addition, the variation in fauna is analyzed over a short period of time, and studies that establish a relationship between diurnal and nocturnal zooplankton are still rare. Thus, although there are taxonomic studies that evaluate organisms from coastal regions, very few analyze the entire zooplankton community specific to tidal flats. As previously mentioned, it is believed that this is directly related to the great difficulties of collecting data in upwelling zones due to the high hydrodynamics that exist, especially on exposed beaches (McLachlan & Brown, 2006).

4. Area of study

The study area is located in the northern sector of the Municipality of Praia Grande, on the Central Coast of the State of São Paulo (Figure 4).

Inserted in the metropolitan region of Baixada Santista, the city of Praia Grande covers 145 km^2 between the coordinates 24°00'/24°05'S and 46°24'/46°35'W, confluencing with the municipalities of Sao Vicente and Mongaguà (Praia Grande, 2015). Throughout the whole of Praia Grande, with the exception of the Xixovà and Itaipù hills, the crystalline bedrock does not reach the ocean waters, thus allowing the development of the extensive and continuous coastline corresponding to 22km of beach (Praia Grande, 2015).

A study point was selected along the municipality's coastline, corresponding to a section of Praia da Aviaçao beach (24°01'S and 46°25'W) located 3.5km from Itaipù hill (the northern end of the coastline).

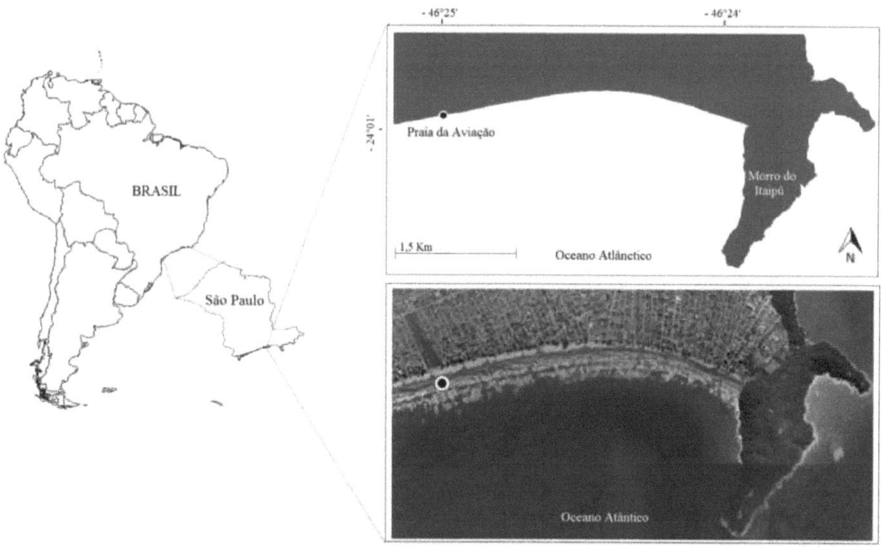

Figure 4: Map of the study area, located in the central area of the São Paulo coast. Source: Modified from Matos *et al.*, 2012; Image source: Google 2013 cartographic data, MapLink. images Cnes/Spot, Digital Globe, Landsat, Terra Metrics.

4.1. Regional Aspects of the Study Area

4.1.1. Terrestrial Geomorphology

The São Paulo coast has geological features linked to the Mesozoic era, through the beginning of geological events that gave rise to the formation of the South Atlantic Ocean (Jurassic-Cretaceous), and later tectono-magmatic events that promoted the uplift of the continental edge and subsidence of the marginal basins (Almeida, 1976; Souza, 1997).

According to Amorim & Oliveira (2009), the Baixada Santista region contains two geomorphological sectors: the Atlantic Plateau Morphostructural Domain, to which the Serra do Mar belongs, and the Coastal Plain Morphostructural Domain.

The Atlantic Plateau Morphostructural Domain has two distinct geoenvironmental units, the Serra do Mar unit and the Residual or Isolated Hills (Amorim, 2007). Both units have a geological formation composed of rocks dating from the Precambrian that have been metamorphized and have acquired high resistance to the action of weathering (Suguio & Martin, 1978). Specifically, the Residual or Isolated Hills, which in the Municipality of Praia Grande are represented by Morro Xixovà and Itaipù, have their formation associated with the tectonic processes that occurred during the Cretaceous-Tertiary, responsible for the fragmentation and isolation of the Serra do Mar towards the ocean basin (IPT, 1981; Sao Paulo, 1997).

The Coastal Plain Morphostructural Domain (Suguio & Martin, 1978) or Coastal Complex (Almeida, 1964; 1976) is made up of continuous areas by the sea, no higher than 70m above sea level and predominantly flat topography, favoring the occurrence of depositional processes over time (Almeida, 1964). The lithology of the Coastal Plain is made up of unconsolidated sands, formed predominantly by Pleistocene and Holocene marine deposits, associated with fluvial and Iagunar deposits (Suguio & Martin, 1978; IPT, 1981). These sediments come from the Cananéia Formation and the Santos Transgression, respectively, associated with sea level oscillations linked to the glacial-eustatic variations of the Quaternary, which gave rise to the marine and fluvial-marine sedimentary deposits in the lower parts of the plain (Suguio & Petri, 1973; Fùlfaro & Suguio, 1974; Suguio & Martin, 1978). According to Souza (1997), these variations in sea level have acted fundamentally as a shaping agent for the coastal plains in the state of Sao Paulo, as well as for the coastline in general.

The current coastline of the state of São Paulo, taking into account its complexity and degree of indentations, is 863.8 km long, with the Baixada Santista region occupying most of this stretch (Lampareli, 1998). According to Souza & Suguio (1996), 430 km of the São Paulo coastline is made up of sandy ocean beaches that have different characteristics along the coast, depending on the geomorphological features of the coastal zone. On the North Coast, the proximity of the Serra do Mar to the ocean means that the regional coastline becomes quite jagged, restricting the extent of the coastal plain and forming features such as inlets, bays and small sheltered beaches (Fùlfaro & Suguio, 1974; Fùlfaro & Coimbra, 1972; Souza, 1997, 2012).

From Baixada Santista, towards the South Coast, the cliffs gradually move away from the coastline, reaching tens of kilometers, and consequently, rocky spurs become scarcer, with the presence of islands, hills and isolated morretes being rare (Fùlfaro & Coimbra, 1972; Melo & Ponçano, 1983; Souza, 1997). This coastal physiography allows for the formation of a straight coastline with a wide sedimentary plain, cut out by large estuarine-laguar systems, and beaches with elongated features exposed to the ocean (Souza, 1997).

The transitional stretch between these two geomorphological features on the São Paulo coast occurs precisely in the Baixada Santista region, which makes the local plateaus highly complex. The northern part of the region is characterized by emergence, with the occurrence of isolated hills and morretes that reach the marine environment and lead to the formation of short beaches and inlets. Towards the south, from the municipality of Praia Grande, the plain is characterized by submergence, with continuous beaches that are less jagged and extensive (Fùlfaro & Ponçano, 1974; Ponçano, 1976).

Thus, the sandy beaches along the coastline of the state of São Paulo have different degrees of exposure, mainly related to the geomorphological characteristics along the São Paulo coastline, which create beaches with different morphological patterns, whether they are more sheltered or exposed. However, according to Souza (1997, 2012), these differences are also related to the predominant peculiarities of the beach over time, such as hydrodynamic, sedimentological and climatic characteristics.

4.1.2. Marine Geomorphology

The formation and evolution of the Brazilian Southeast Continental Shelf is directly related to the geological events responsible for the opening up of the South Atlantic Ocean and the tectonic events that promoted the uplift of the Serra do Mar and subsidence of the Santos Basin (Almeida, 1976; Souza, 1997).

The continental shelf of the State of São Paulo, formed after the geological events mentioned above, covers a width of between 73 and 231 km, where the depth of the shelf break is located between 120 and 180m, with the portion of Santos Bay being more extensive when compared to adjacent areas (Rodrigues *et al.*, 2003). The shelf can be divided into different regions: inner, middle and outer (Nittrouer & Wright, 1994). According to the authors, the inner portion of the shelf is associated with coastal zones where hydrodynamics are more intense due to the shallower depths, while the middle shelf is a transitional region where there is a decrease in the intensity and frequency of the waves under the seabed and, finally, the outer region is deeper and reflects a strong decrease in wave action under the seabed.

The chemical composition of the surface sediments of Sao Paulo's continental shelf has been described in various studies, including the study by Alves & Ponzi (1984), which characterizes the inner and middle shelf as consisting mainly of terrigenous sediments and the outer shelf as having a higher percentage of carbonaceous sediments. With regard to the granulometric pattern of the surface sediments on the state's continental shelf, this is described by Folk & Ward (1957) and Furtado & Mahiques (1990) as consisting of fine and very fine sands.

4.1.3. Oceanography

In order to understand oceanographic phenomena, it must be considered that the São Paulo coast, located on the Southeast Continental Shelf, is subject to the influence of large-scale oceanographic events occurring in this region (Fontes, 1995). According to the author, these events respond to climatic phenomena related mainly to the effects of wind dynamics linked to circulation of maritime origin and the movement of frontal systems. In addition, the region is also influenced by regional water masses, which for the state of São Paulo include, according to Fontes (2010), Coastal Water (CA), originating from the interaction of continental waters with marine waters; Central Water of the South Atlantic (ACAS), which originates in the subsurface, is colder and richer in nutrients, and eventually reaches the surface or close to it (during periods of intensified winds from the eastern quadrant); and Tropical Water (AT), which is warmer and poorer in nutrients, coming from the Brazil Current, which eventually forms intrusions on the inner shelf, reaching the beaches.

The São Paulo coast is subject to a semi-diurnal micrometeorological regime, which results in a physical process that has little influence on beach environments or the modification of currents (Mesquita, 1995). However, according to Harari & Camargo (1990), this astronomical component can be intensified in specific situations, such as the entry of cold fronts, oscillating in extreme conditions between +1.4m and -1.8m (Tommasi, 1979). Sea level variations, on a monthly and daily scale, are relatively the same for the entire coast of the state, not exceeding 1m, according to Mesquita (1995). In this way, São Paulo's beaches can be characterized as wave-dominated, since the tide acts secondarily (Souza, 1997).

According to data collected by Tessler *et al.* (2006) and Souza (2012), the height of waves generated in the open ocean varies between 0.5 and 2m, with periods between 3 and 30 seconds and intervals between 9 and 11 seconds. However, higher energy waves from the SW, S and SE quadrants can reach heights of more than 5m when there is an association between frontal systems and extratropical cyclones (Fonzar, 1994; Satyamurty *et al.*, 1998). According to the authors, these atmospheric disturbances are responsible for hangover events and occur most frequently between April and September.

In addition to the action of the winds on the Continental Shelf, which are responsible for the formation of waves in the open ocean, local winds can also generate waves, but these are not considered to be important agents in modifying the waves that hit the coast (Tessler, 1988; Bomtempo, 1991; Souza, 1997; 2012).

4.1.4. Weather

In general, the climate of South America can be characterized, according to Satyamurty *et al.* (1998), as a mixture of tropical, subtropical and extratropical climates. Its atmospheric circulation is influenced by both the Atlantic Ocean and the Pacific Ocean, due to its location between the two oceans (Wainer & Taschetto, 2006).

The atmospheric circulation in the southern part of South America is influenced by the action of two anticyclones, the Atlantic Tropical Anticyclone (ATA) and the Atlantic Polar Anticyclone (APA) (Fonzar, 1994). According to Sant'Anna Neto (1994), the ATA gives rise to the Atlantic Tropical Mass, which is active all year round on the São Paulo coast and is responsible for generating winds called E and NE trade winds. The APA forms the Atlantic Polar Mass which moves frontally towards the coast and is responsible for the occurrence of cold fronts which alter the direction and intensity of the prevailing winds in the region, as well as atmospheric pressure and precipitation rates (Nimer, 1979; Fonzar, 1994; Sant'Anna Neto, 1994; Satyamurty *et al.*, 1998). Therefore, on the São Paulo coast, the climate is directly influenced by tropical and polar masses, with the action of the polar mass and, consequently, the occurrence of cold fronts being more rigorous and frequent in winter (Sant'Anna Neto, 1994).

According to Alvares *et al.* (2013), practically the entire coastline of the state of Sao Paulo, including the southern region, has the characteristics of a Tropical Humid Climate (Af), according to Koppen's international climate classification. The areas under the influence of this climate regime have no dry season and average temperatures ≥18°C (coldest month) and ≥22°C (hottest month) (Rolim, 2007). The escarpments of the Serra do Mar directly influence the formation of orographic rainfall (Conti &

Furlan, 2011) and, consequently, the high annual rainfall rates (3,000mm) recorded in the central region of the state (Alvares *et al.*, 2013).

This climatic characterization is determined, on the São Paulo coast, mainly by the geomorphology of the coast, composed of an irregular topography with mountains, hills, valleys and plains, capable of directly influencing the atmospheric circulation at low altitudes (Monteiro, 1973; Sant'Anna Neto, 1994; Amorim & Oliveira, 2007).

According to Monteiro (1973) another factor that influences the climate in the region is the Brazilian ocean current, as this warm water current increases the amount of water available in the system through evaporation. However, due to the predominance of the Atlantic Tropical Mass, the weather remains stable, except in the winter months when the region is more subject to strong cooling due to the influence of the Atlantic Polar Mass.

Specifically for the Baixada Santista region, there is very little literature on climatic aspects, and the climate is associated with a series of old publications. The climatological panorama of the Baixada Santista, for two decades (1980 to 2004), was described by Harari *et al.* (2008) where the authors characterized the region based on the climate model developed by the National Center for Environmental Prediction / National Center for Atmospheric Research (NCEP/NCAR). ±This study showed rainfall rates of between 1734.45 and 3027.50mm/year and significant seasonal variations in air temperature, with averages of 17.33 ± 3.89°C in July and 21.77 ± 1.89°C in February. As far as atmospheric circulation is concerned, the study shows a predominance of winds from the eastern quadrant. The annual rainfall pattern is highly concentrated in the summer season, due to the presence of humid air masses that move towards the Serra do Mar, generating high cloudiness and rainfall (Sao Paulo, 2010). The least rainy period occurs in the opposite direction, but there are often instabilities related to the entry of cold fronts responsible for heavy rainfall in just a few days (Harari *et al.*, 2008; Sao Paulo, 2010).

4.2. Local Aspects of the Study Area

The Praia Grande coastline, together with the municipalities of Mongaguà, Itanhaém and Peruibe, forms one of the longest beach arcs on the São Paulo coast, approximately 70km long (Souza, 2012) and, according to the author, the coastal plain and continental shelf of the region are broad and have low topographic gradients.

The Praia Grande coastline is bordered to the north by Itaipù Hill and to the south by the territorial limit with the city of Mongaguà. In the far north there is an abrupt contact between the ocean and Morro de Itaipù, characterizing a spur that runs into the ocean, where rocky coasts form (Guerra & Guerra, 2005) with a high bathymetric gradient (Sao Paulo, 1997). This rock formation is exposed to the action of waves and acts as an obstacle to the propagation of wave trains that reach the coast (Sao Paulo, 1997).

The beach strip is essentially made up of fine- to very fine-grained sandy sediment, this material being very well sorted due to the small dispersion between the granulometric values (Souza, 1997, 2012). This characteristic is mainly due to the process of beach formation from the reworking of ancient marine sediments (Tesller, 1988; Guedes, 2009; Souza, 2012).

According to Souza (1997) and Farinaccio (2008), the slope and width of the beach strip vary greatly along the coastline, with the northern portion having the least slope and the greatest width (150m), while this pattern changes towards the south, where there is the greatest slope and the shortest stretch of beach strip (80m). The coastline is oriented NE-SW and is completely exposed to the S-SSE swell system (Souza, 1997).

With regard to the morphodynamic pattern, the low slope characteristics (average slope of 2°, which can reach 4° at some points), extensive surf zone and constant exposure to high-energy waves mean that the Praia Grande beach strip has dissipative beach features (Souza, 1997) (Figure 5). However, the morphodynamic pattern can change depending on weather conditions and local hydrodynamic variations (Souza, 1997, 2001; Souza & Suguio, 1996, 2003).

Figure 5: Local images of the study area - Praia da Aviaçâo (Praia Grande/SP). A) View of the northern portion; C) Front view of the beach.

Longitudinal beach currents are common on the Praia Grande coast, flowing in two directions (NE and SW), although there is a slight predominance of SW currents, towards the city of Mongaguâ (Souza, 1997). According to Bomtempo (1991) the longitudinal current flows more frequently to the SW during calm periods and to the NE during cold fronts.

Currently, the municipality's waterfront is not directly influenced by continental watercourses of natural origin, mainly due to the intense process of urbanization which, among other consequences, has significantly altered the surface drainage pattern of the coastal plain. However, over time, there has been a progressive increase in artificial river channels on the beach strip, which drain towards the sea and consequently cause the input of fresh water and organic matter to the local marine ecosystem (Souza & Cunha, 2009, 2012) (Figure 6 and 7).

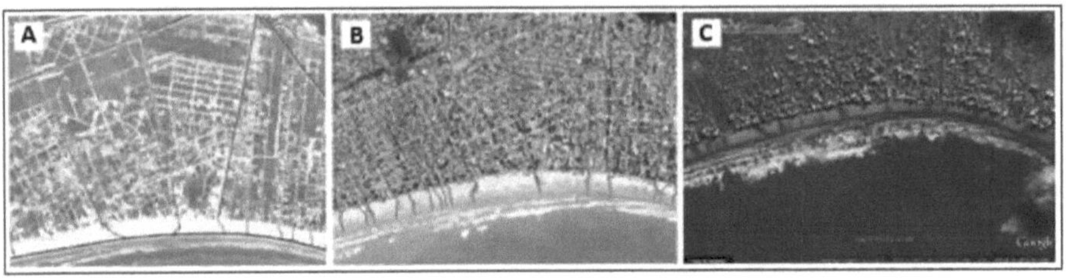

Figure 6: Aerial photographs showing the emergence of new pluvial-fluvial channels along the time, on the Praia Grande waterfront. A) Image from 1962, illustrating the presence of a few canals; B) Image from 2000 and C) Image from 2015. Source: Modified from Souza & Cunha (2009, 2012) and cartographic data.

Google 2015, MapLink. images Cnes/Spot, Digital Globe, Landsat, Terra Metrics.

Figure 7: Images of rainfall-runoff in the study area. A) View of the river discharge point;

B) and C) View of the runoff on the beach strip. Source: Vaske Jr. (2014) (personal collection)

5. Methods

5.1. Zooplankton Sampling and Obtaining Environmental Variables

Data was collected at a point made up of three replicates equidistant every 100m, located within the internal surf zone on a stretch of Praia da Aviaçao, in the municipality of Praia Grande - SP. The selected point was sampled bi-monthly for two consecutive years, from June 2013 to April 2015, during the day and night (Figure 8). During the sampling period, it was possible to obtain data in all seasons of the year with conditions of ordinary high tide and low tide of syzygy and quadrature.

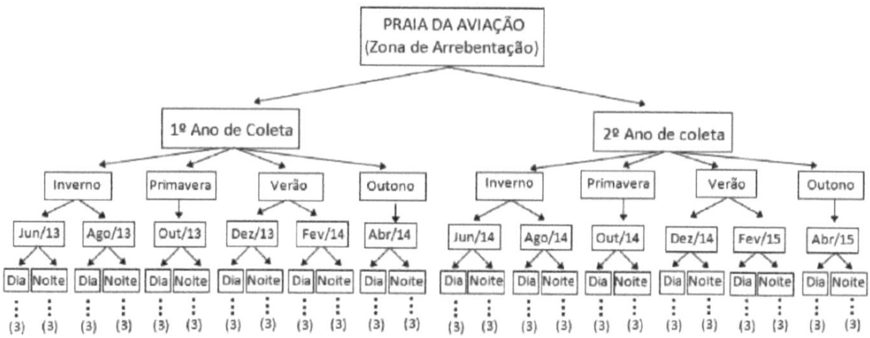

Figure 8: Sampling design for the collection of zooplankton fauna in the internal surf zone of Praia da Aviaçâo (Praia Grande - SP). ( )= number of sampling replicates.

The zooplankton fauna was collected by continuous surface trawling, horizontally and parallel to the beach, using a cylindrical-conical plankton net with a diameter of 50 cm, a length of 1.4m and a mesh size of 180μm. The trawling was carried out by two people walking 100m apart for three minutes at a depth of 1m (Figure 9a, 9b, 9c). While still in the field, the captured organisms were anaesthetized with magnesium chloride (7.5%) to prevent the organisms from contracting, expelling their intestinal contents and losing their egg sacs. Formaldehyde buffered with sodium tetraborate at a final concentration of 4% was used to preserve and fix the samples, since the acidic action of formaldehyde can cause damage to the organisms over time (Haney & Hall, 1973). Finally, the sample contents were stored in plastic bottles with a capacity of 250 ml (Figure 9d).

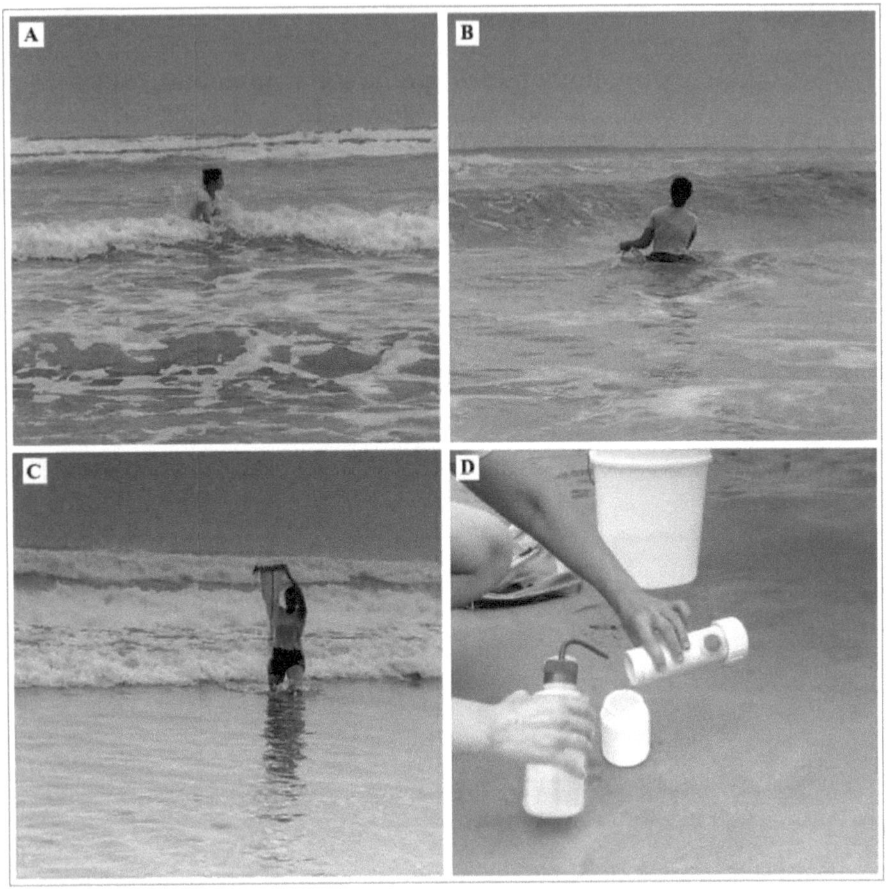

Figure 9: Collection of zooplankton fauna in the surf zone. A) and B) Surface trawl, horizontal and parallel to the beach; C) Removal of the water from the collection net; D) Storage of the samples.

At the time of collection, abiotic data on *water* temperature and salinity were obtained, as well as quantification of the number of incident waves, observations of the state of sea agitation, local wind and cloud cover, tide and lunar cycle.

Water temperature and salinity were obtained using a mercury thermometer and a refractometer, respectively. The Beaufort scale was used to classify the local wind and the aspect of the sea. Cloud cover was determined by visual observation, by dividing the (imaginary) sky vault into eight quadrants and adding up the clouded parts, according to the standards defined by the World Meteorological Organization (Echer *et al.*, 2006). Tide and lunar cycle data were obtained from the Brazilian Navy's Directorate of Hydrology and Navigation (DHN) from the National Oceanographic Data Bank (BNDO).

5.2. Sample Processing and Fauna Analysis Zooplankton

In the laboratory, due to the high number of organisms, the samples were reduced using a "Folsom Plankton Splitter" (Omori & Ikeda, 1984) to obtain sub-samples (Boltovskoy, 1981), varying the size of the fractions (1/4 to 1/256) depending on the concentration of organisms (Figure 10). The number of individuals recorded was then multiplied by the subsampling factor in order to estimate the number of organisms collected.

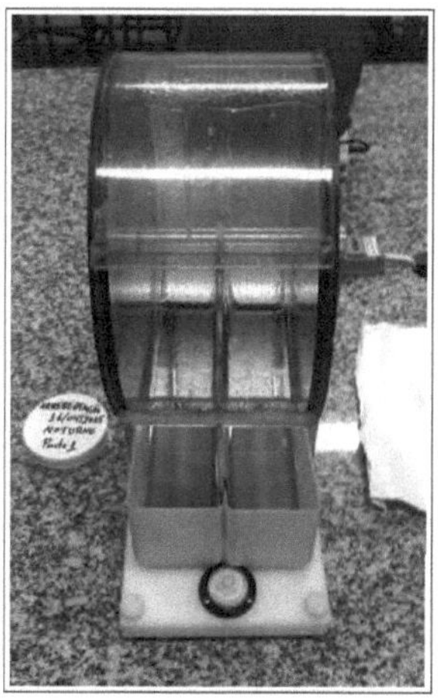

Figure 10: Fractionation of the samples due to the high abundance of zooplankton organisms captured.

The taxa were sorted, identified and quantified under a binocular stereomicroscope (Zeiss). For the qualitative analysis of the fauna, the guides compiled by Smith & Johnson (1996), Boltovskoy (1999), Johnson & Allen (2005) and Suthers & Rissik (2009) were used. The identification of the zooplankton was carried out at different taxonomic levels, always looking for the best refinement, with some being classified at genus and species level, but the majority being recognized at order and class level. The identification of the organisms was validated and updated in terms of nomenclature and authorship according to the *WoRMS* (*World Register of Marine Species*) database. The taxons identified were listed in large groups for better graphic visualization of the general composition.

The zooplankton fauna present in the shoaling zone was classified according to its frequency of capture, and termed "resident", "migratory" or "sporadic", according to Clarck (1996a). According to this author, "resident" organisms are those captured practically every month of the sample, "migratory" refers to taxa collected at certain times of the year or in the same months of subsequent years, which can reach the rapture zone voluntarily or involuntarily, and those classified as "sporadic" refer to organisms whose occurrence does not follow a set pattern. In this study, the zooplankton, holoplankton, meroplankton and bentoplankton fauna were also grouped according to their life habits. The organisms were also classified according to their stage of development (egg, larva, juvenile and adult). The benthic fauna captured in the adult stage was also organized according to its habitat (benthic sandy substrate, benthic rocky substrate and bentoplankton). The bentoplankton group is made up of organisms that are linked to the sandy substrate, but which have the characteristic and capacity to migrate to the plankton at certain times of the day (McLachlan & Brown, 2006). However, adult organisms with an exclusively benthic habit were not included in the analysis, as it was considered that they do not belong to the zooplankton fauna and are only present in the plankton as a result of the high hydrodynamics of the upwelling zone.

The abundance of the zooplankton fauna (org.m-^{3}) was calculated using the formula proposed by Newell & Newell (1963), which considers the total number of individuals present in the sample (estimated in the subsample), divided by the volume of water filtered (m3) using the formula: n° org.m3 = n / v (m3); where n represents the number of organisms in the sample; and v the volume of water filtered by the net. The volume of water released was calculated as a function of the net's opening area (m^2), speed (m/s) and drag time (sec.) using the formula: V (m3) = A x v x T; where A corresponds to the net opening *(πr^2)*; v represents drag speed (distance divided by drag time); and

T corresponds to drag time.

The percentage contribution of the zooplankton groups was obtained by dividing the abundance data for each taxonomic group by the total abundance of the fauna captured. The frequency of occurrence was determined by dividing the number of samples in which the taxon was recorded by the total number of collections made, and the result was presented as a percentage, so the taxonomic groups were classified as very frequent ($n > 75\%$), frequent ($50\% < n \leq 75\%$), sporadic ($25\% < n \leq 50\%$) or rare ($n \leq 25\%$).

Diversity was obtained using the *Shannon* index, considered the most recommended index for ecological studies (Krebs, 1989), which is based on the proportion of species *i* in the sample *(pi),* which can be considered as the probability that an individual sampled at random belongs to species *i*: H' (bits.ind-^{1}) = $-\sum$ (pi) (log2 pi); where: H'= Shannon's diversity index; pi= proportion of individuals of species *i* in the sample. The richness of the fauna was determined by the number of taxa captured in each sample. Equitability (J'), capable of indicating the uniformity between the abundance of the identified taxa, was measured using the formula proposed by Pielou (1977): $J' = H' / H'$ (maximum); where: H' corresponds to *Shannon*'s index ; H' (maximum) = log(*S*) represents the maximum diversity that can be observed if all species present equal abundance, where *S* refers to the total number of taxons sampled. For the diversity and evenness calculations, the total abundance data of the identified taxa was used; however, the dominant group (Copepoda) was not considered due to its high abundance and possible interference with the result.

5.3. Data Processing and Statistical Analysis

In the statistical analysis, the variables examined were the zooplankton community (holo, mero and bentoplankton), time of day, month of sampling and temperature (°C) of the sea water in the surf zone, all analyzed in relation to the total abundance of the fauna. To verify the normality and homogeneity of the variables, they were analyzed using the *Shapiro-Wilk* and *Cochran* tests, respectively. In the case of non-parametrically distributed variables, the *Generalized Linear Model (GLM)* was used to determine the contribution of the factors in relation to the variation in zooplankton abundance observed in the upwelling zone. The analyses were first carried out between the four variables (C = community; M = sampling month; P = sampling period; T = water temperature), always relating them to the total abundance data of the fauna. The significant variables were separated in order to create a model representing the factors that really influenced the abundance of the fauna. A new *GLM* was then run with only the significant variables. These analyses were carried out using the *Statgraphics* program.

The multivariate analyses were based on the average abundance data of the zooplankton fauna recorded at the climatic stations during the two sampling years. The *Non Multi Dimensional Scaling (NMDS)* technique was used to recognize a certain degree of similarity among the sampling campaigns so that they could be grouped according to seasonal variation. Similarity analysis by hierarchical agglomeration (*cluster*) was used to verify the formation of groupings. This analysis was based on the Bray-Curtis index (Bray-Curtis, 1957), with the similarity matrices calculated in "Q" mode. The groupings generated by the *cluster* analysis with a maximum linkage distance equal to 85% were plotted on the *NMDS* to better visualize the data distribution patterns.

Multivariate analyses were carried out using *PRIMER* 6.0 *software* (*Plymouth Routines in Multivariate Ecological Research*, 2006).

6. Results

6.1. Environmental variables

The weather during the period analyzed, in terms of wind patterns, ranged from a light breeze to a weak wind, according to the Beaufort scale. Most of the sampling took place under overcast skies and in the crescent moon phase. With regard to the local hydrodynamic characteristics, in most of the samples, the sea had small swells associated with a low number of waves (Table 1).

Table 1: General weather and sea conditions during the sampling period (BM = low tide; D = daytime; N = nighttime; PM = high tide).

Year	Month		Weather				Sea	
			Wind*	Coverage of Clouds	Stage of Moon	Tide	Sea View*	Number Wave
	Jun.	D N	Light breeze	Slightly overcast Undercover	Growing	PM BM	Small ripples	5
		D	Light breeze				Small ripples	5
2013	**Aug.**	N	Weak	Almost overcast	Growing	BM	Ripples and some sheep	7
	Oct.	D N	Light breeze	Undercover	Growing	PM BM	Small ripples	8 4
	Dec.	D N	Weak	Undercover	Growing	BM	Ripples and some sheep	6 10
	Feb.	D N	Light breeze	Half overcast Undercover	Growing	BM PM	Small ripples	5 4
	Apr.	D	Weak	Undercover	Waning	PM	Ripples and some sheep	7
		N	Light breeze			BM	Small ripples	6
	Jun.	D	Weak	Undercover	Growing	PM	Ripples and some sheep	6
		N	Light breeze	Half overcast		BM	Small ripples	5
2014		D	Light breeze	Slightly overcast			Small ripples	3
	Aug.	N	Weak	Undercover	Full	BM	Ripples and some sheep	7
	Oct.	D N	Light breeze	Undercover	Waning	BM PM	Small ripples	5
	Dec.	D N	Light breeze	Almost overcast	Full	BM	Small ripples	4 5
	Feb.	D N	Light breeze	Slightly overcast Half overcast	Growing	PM PM	Small ripples	5
2015	**Apr.**	D N	Light breeze	Undercover Almost overcast	Waning	BM	Small ripples	4 6

* Based on the Beaufort scale

The highest water temperature values were recorded in summer and ranged from 28.5°C (day) to 30°C (night), both in February/2015. The lowest temperatures occurred during the winter, ranging from 18.7°C (day) in August/2013 to 16.7°C (night) in

August/2014.

Salinity showed small variations over the period analyzed, with a minimum of 33 and a maximum of 35.4 in April and October 2014, respectively (Figure 11). However, the salinity averages in relation to the time of day were very similar, ranging from 34.4 (day) to 34 (night).

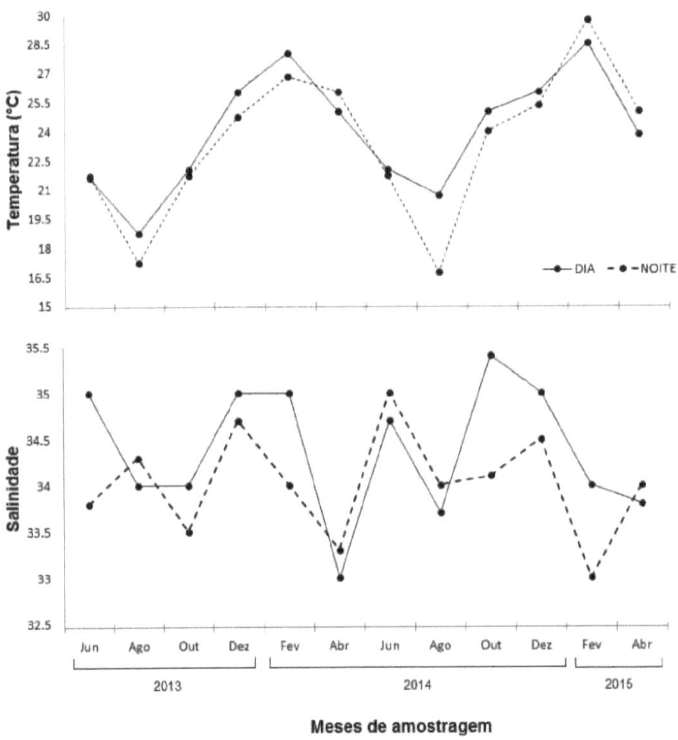

Figure 11: Abiotic data from the water in the inner surf zone of Praia da Aviaçao.

6.2. Zooplankton community

6.2.1. Fauna composition

In total, 116 taxons were recorded from the samples taken over two years of analysis, during the day and night (Table 2).

Following the classification according to life habit, holoplankton represented 36.2%

(N=42), meroplankton 14.7% (N=17), benthic animals from sandy and rocky substrates 26.7% (N=31) and 12.1% (N=14), respectively, and bentoplankton 6% (N=7) of the taxa captured during the study. Organisms in the meroplanktonic and adult stages of the same taxonomic group were also collected, totaling 4.3% (N=5) (Figure 12).

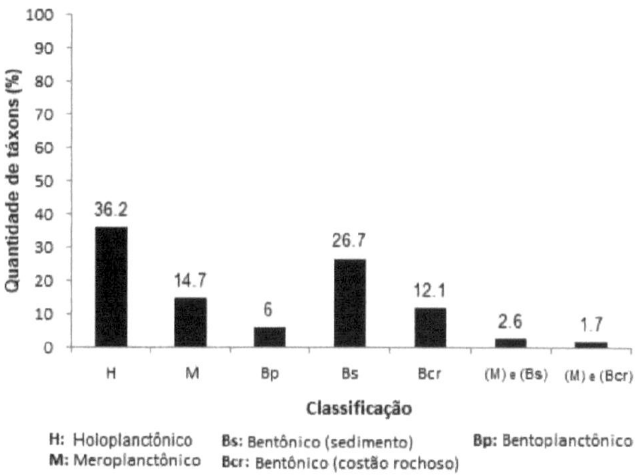

Figure 12: Representation of the number of taxons recorded, according to life habit, during the study period.

The holoplankton fauna was composed more abundantly of Copepoda (Calanoida), Cladocera (Podonidae) and Tunicata (Oikopleuridae), as well as Chaetognatha (Sagittidae), Cnidaria (Hydrozoa) and Actinopteri (Engraulidae). Tintinnina, Foraminifera, Ctenophora, Cubozoa and Scyphozoa were less common.

Table 2: List of taxons present in the inner surf zone of Praia da Aviaçà, caught bimonthly over two years during the day and night. **Est.** = Life stage; Ad. = Adult; Juv. = Juvenile; La. = Larvae; PI. = Post-larva; Meg. = Megalopa; Nau. = Nauplio; **Cla.** = Classification according to life history; H. = Holoplanktonic; M. = Meroplanktonic; Bp. = Bentoplanktonic; Bs. = Benthic sandy substrate; Ber. = Benthic rocky shore; N.I. = Not identified.

Táxon	Est.	Clas.
Tintinnina	Ad	H
Foraminifera		
Lagenida		
Vaginulinidae		
Lenticulina sp.	Ad	Bs
Littorinimorpha		
Ranellidae		
Gelagna sp.	Ad	Bs
Miliolida		
Hauerinidae		
Quinqueloculina sp.	Ad	Bs
Pyrgoella sp.	Ad	Bs
Rotaliida		
Anomalinidae		
Hanzawaia sp.	Ad	Bs
Bolivinellidae	Ad	Bs
Bolivinitidae	Ad	Bs
Buliminidae	Ad	Bs
Cibicididae	Ad	Bs
Discorbinellidae	Ad	Bs
Elphidiidae	Ad	Bs
Globigerinidae		
Globigerinoides ruber	Ad	H
Globigerina bulloides	Ad	H
Orbulina universa	Ad	H
Nonionidae		
Nonionella atlantica	Ad	Bs
Stainforthiidae	Ad	Bs
Uvigerinidae	Ad	Bs
Foraminifera N.I.	Ad	H
Foraminifera N.I.	Ad	Bs
Cnidaria		
Hydrozoa		
Anthoathecata		
Hydractiniidae		
Hydractinia sp.	Ad	Bcr
Hydridae		
Hydra sp.	Ad	Bcr
Tubulariidae	Juv/Ad	M/Bcr
Anthoathecata N.I.	Ad	Bcr
Cnidaria		
Hydrozoa		
Leptothecata		
Aglaopheniidae		
Aglaophenia sp.	Ad	Bcr
Campanulariidae		
Obelia sp.	Juv/Ad	M/Bcr
Phialellidae	Ad	Bcr
Sertulariidae	Ad	Bcr
Dynamena sp.	Ad	Bcr
Sertularia sp.	Ad	Bcr
Siphonophorae		
Abylidae	Ad	H
Apolemiadae	Ad	H
Diphyidae	Ad	H
Physaliidae		
Physalia sp.	Juv	H
Rhizophysidae		
Rhizophysa sp.	Juv	H
Trachymedusae		
Geryoniidae		
Liriope tetraphylla	Ad	H
Hidrozoa N.I.	Juv	M
Cubozoa	Ad	H
Scyphozoa	Ad	H
Anthozoa		
Pennatulacea		
Renillidae	Ad	Bs
Ctenophora	Juv	H
Bryozoa	La	M
Nemertea	Ad	Bs
Platyhelminthes		
Turbellaria	Ad	Bs
Sipuncula		
Sipunculidae		
Xenosiphon sp.	Ad	Bs
Sipuncula N.I.	Ad	Bs
Mollusca		
Gastropoda	La	M
Bivalvia	La	M
Polychaeta		
Phyllodocida		
Chrysopetalidae	La	M
Glyceridae	La	M
Hesionidae	Ad	Bs
Nereididae	La/Ad	M/Bs
Phyllodocidae	La	M
Syllidae	La	M
Myrianida sp.	Ad	Bs
Sabellida		
Oweniidae	La	M
Spionida		
Spionidae	La/Ad	M/Bs
Magelonidae	La	M
Terebellida		
Pectinariidae	La	M
Terebellidae	La	M
Polychaeta N.I.	La/Ad	M/Bs
Arthopoda		
Copepoda		
Calanoida	Ad	H
Cyclopoida	Ad	H
Harpacticoida	Ad	H
Poecilostomatoida	Ad	H
Siphonostomatoida		
Caligidae		
Caligus sp.	Ad	
Copepoda	Nau	H
Cirripedia	Nau/La	M
Cladocera		
Podonidae		
Evadne sp.	Ad	H
Pseudevadne sp.	Ad	H
Podonidae N.I.	Ad	H
Sididae		
Penilia sp.	Ad	H
Cumacea		
Leuconidae	Ad	Bs
Pseudocumatidae	Ad	Bs
Arthopoda		
Euphausiacea	La	M
Amphipoda		
Gammaridea	Ad	Bp
Hyperiidea		
Hyperiidae		
Hyperoche sp.	Ad	Bp
Isopoda		
Bopyridae	Ad	Bp
Idoteidae		
Edotia sp.	Ad	
Isopoda N.I.	Ad	Bp
Mysida		Bp
Mysidae		
Bowmaniella brasiliensis	Juv/Ad	Bp
Metamysidopsis elongata	Juv/Ad	Bp
Ostracoda	Ad	Bs
Pycnogonida		
Phoxichilidiidae		
Anoplodactylus sp.	Ad	Bcr
Stomatopoda	Ad	Bs
Decapoda		
Brachyura	Zoea/Meg	M
Portunidae	Ad	Bs
Brachyura N.I.	Ad	Bs
Caridea	Zoea	M
Penaeidae	Pl	M
Penaeus sp.	Ad	Bs
Thalassinidae	Zoea	M
Luciferidae		
Lucifer sp.	Ad	H
Hippidae		
Emerita sp.	Juv	Bs
Echinodermata		
Asteroidea	La	M
Echinoidea	La/Juv	M
Holothuroidea	La	M
Invertebrado	Ovo	
Chaetognatha		
Sagitta sp.	Ad	H
Chordata		
Tunicata		
Appendicularia		
Oikopleuridae		
Oikopleura sp.	Ad	H
Ascidiacea	La	M
Thaliacea		
Salpidae		
Thalia sp.	Ad	H
Doliolida	Ad	H
Cephalochordata		
Leptocardii		
Branchiostomatidae	Ad	Bs
Vertebrata		
Actinopteri	Ovo/La	H
Atherinopsidae		
Atherinella blackburni	Juv	H
Sciaenidae		
Menticirrhus littoralis	Juv	H
Carangidae		
Oligoplites sp.	Juv	H
Haemulidae		
Conodon nobilis	Juv	H
Mugilidae		
Mugil brevirostris	Juv	H
Mugil curema	Juv	H
Cynoglossidae		
Symphurus sp.	Juv	H
Linguado sp.1	Juv	H
Linguado sp.2	Juv	H
Linguado sp.3	Juv	H
Engraulidae		
Anchoviella lepidentostole	Juv	H
Exocoetidae	Juv	H
Gobiidae	Juv	H

The meroplankton organisms were best represented by larval stages of Gastropoda and Bivalvia, Polychaeta (Spionidae), Echinodermata (Echinoidea), Cirripedia, Decapods (Brachyura and Thalassinidae) and Euphausiacea. Other members of the meroplankton, Bryozoa and Tunicata, occurred less frequently.

Among the benthic organisms inhabiting sandy substrates, Foraminifera, Sipuncula, Polychaeta, Ostracoda, Stomatopoda, Decapoda, Cumacea and Chephalochordata were collected. As for benthic animals from rocky substrates, Cnidaria were the main ones captured, followed by Pycnogonida.

The benthoplanktonic individuals frequently caught were microcrustaceans from the Amphipoda, Isopoda and Mysida groups.

In relation to the pattern of occupation of the upwelling zone by the zooplankton fauna identified in this study, following the classification proposed by Clark *et al.* (1996), where organisms are named according to their frequency of capture, 27 taxa were recorded as resident in the upwelling zone of Praia da Aviaçao, of which 48.1% (n = 13) were meroplanktonic, 33.3% (n = 9) holoplanktonic and 18.5% (n = 5) bentoplanktonic. Only two taxa were classified as migratory, one of which was meroplanktonic, present in the surf zone in August, October and April, and the other bentoplanktonic, caught only in August. Another twelve taxa showed less obvious patterns of occurrence, ten of which were holoplanktonic and two meroplanktonic. The rest of the fauna, 31 taxa, occurred sporadically in the surf zone (Annex 2).

6.2.2. Frequency of Occurrence

In the first year of sampling, 13 taxonomic groups were caught very frequently during the day, three of which were recorded frequently, six sporadically and only one rarely. During the night, the very frequent fauna was the same as that captured during the day, but two taxonomic groups were added, five groups were frequent and five were sporadic (Figure 13a).

In relation to the second year, twelve groups were very frequent during the day, four were recorded as occurring frequently, seven sporadically and one rarely. At night, a high frequency of occurrence was recorded for fourteen taxonomic groups, three frequent, seven sporadic and two rarely recorded. In this second year of sampling, three taxonomic groups were not recorded during the day (Tintinnina, Luciferidae and Ascidiacea), while only one was not captured during the night (Ascidiacea) (Figure 13b).

In general, of the 27 taxonomic groups identified throughout the study during the day, eleven were caught very frequently (Copepoda, Chaetognatha, Tunicata, Actinopteri - egg, Cnidaria - meroplanktonic, Mollusca, Polychaeta, Cirripedia, Brachyura, Amphipoda and Isopoda), seven frequently recorded (Foraminifera, Cnidaria - holoplanktonic, Cladocera, Bryozoa, Euphausiacea, Echinodermata and Mysida), three sporadically caught (Actinopteri - larva, Caridea and Penneidae) and four rarely occurring (Ctenophora, Actinopteri - juvenile, Thalassinidae and Ascidiacea). At night, twelve were very frequent (Copepoda, Chaetognatha, Tunicata, Actinopteri - egg, Cnidaria - meroplanktonic, Mollusca, Polychaeta, Cirripedia, Brachyura, Penneidae, Amphipoda and Mysida), six groups were frequent (Foraminifera, Cnidaria - holoplanktonic, Cladocera, Actinopteri - juvenile, Echinodermata and Isopoda), five recorded sporadically (Actinopteri - larva, Bryozoa, Caridea, Thalassinidae and Euphausiacea) and four rarely recorded (Tintinnina, Ctenophora, Luciferidae and Ascidiacea) (Figure 13c).

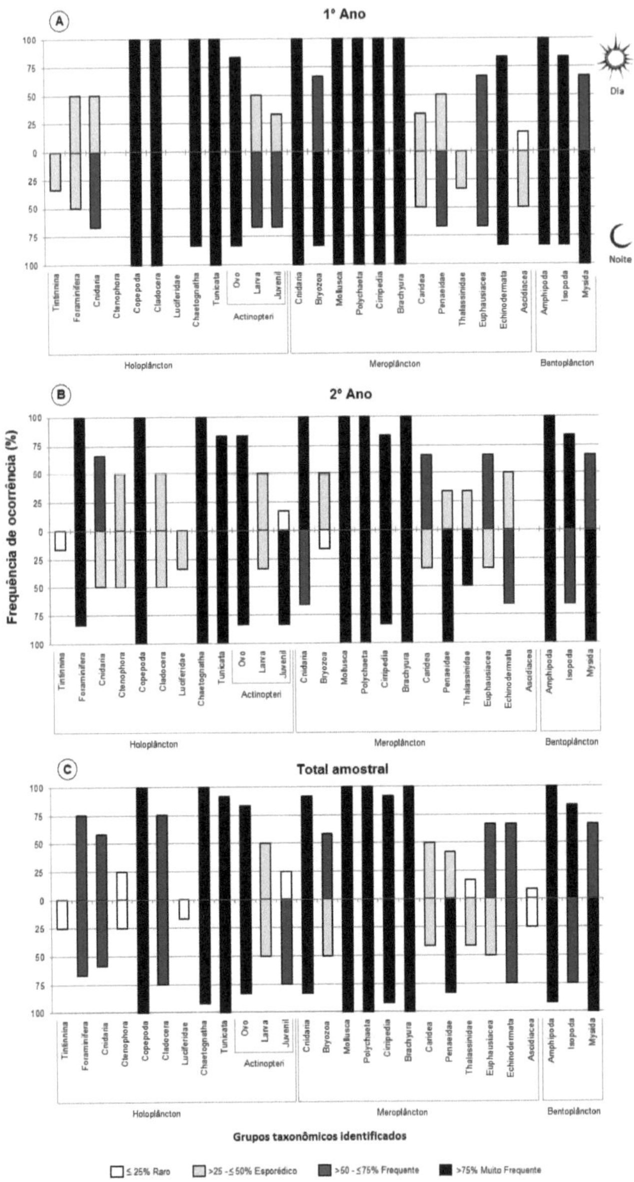

Figure 13: Frequency of occurrence of taxonomic groups caught during the study, day and night.

6.2.3. Percentage Contribution

Throughout the study, including both sampling periods, the Copepoda group contributed the highest percentage (84%) of the total zooplankton fauna caught. Among the holoplankton, in addition to Copepoda, the groups with the greatest contribution were Cladocera (4%) and Tunicata (2%). The meroplanktonic forms with the highest contribution were the Mollusca group (3%) and Echinodermata (3%) and, in relation to the bentoplanktonic forms, the Mysida group (2%) showed the highest percentage of contribution among the other organisms. The total of the other groups caught corresponded to 2% of all the zooplankton fauna recorded throughout the study (Figure 14).

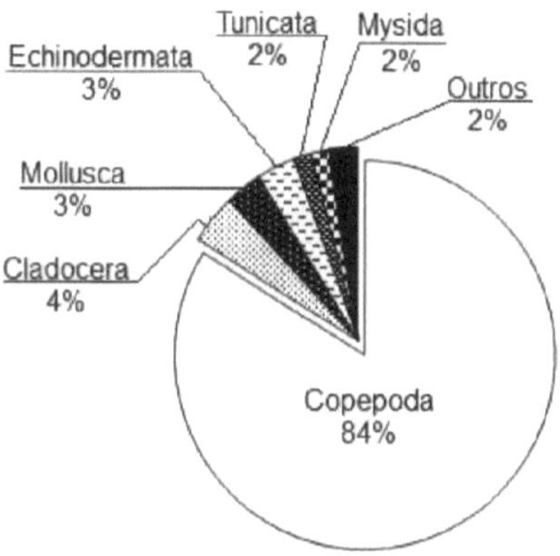

Figura 14: Percentage contribution of the main taxonomic groups sampled in the surf zone of Praia da Aviaçâo.

The monthly analysis of the variations in the percentage contribution of the most abundant taxa during the sampling periods (day and night) is shown in figure 15. In this analysis, the Copepoda group, which was the most abundant, was removed in an attempt to provide better visualization and representativeness of the other groups evaluated.

It was thus observed that the holoplankton community in the first year was dominated by tunicates, represented by the Appendicularia and Thaliacea classes, marked by a greater abundance of the *Oikopleura* and *Thalia* genera when compared to the second sampling year. The ketognaths of the genus *Sagitta* were also representative in the holoplankton community, with their contribution being more noticeable in the second sampling year. However, other taxonomic groups did show a high contribution, such as Actinopteri, Cladocera composed more abundantly of the genera *Penilia* and *Pseudevadne*, and the Foraminifera group with a higher abundance of *Globigerina* and *Globigerinoides* (Figure 15a).

Among the meroplankton fauna, molluscs made up of bivalve and gastropod larvae were the most dominant group (>50%) in both collection periods, in most of the months analyzed, except for the daytime period in December/2013 and 2014, the nighttime period in April/2014 and both periods in April/2015, when echinoderms (echinopluteus) dominated the meroplankton fauna. The larvae of polychaetes, cirripods and decapods were also recorded, but with less dominance (Figure 15b).

The benthoplanktonic community was best represented by misidaceans, which was the most expressive group in sixteen of the twenty-four samples taken during the period analyzed. The Amphipoda group was also very expressive during the day, for example in October/2013, February, August and October/2014 and April/2015. Isopods showed high contribution rates in samples from the daytime period in June/2013 and at night in February/15 (Figure 15c).

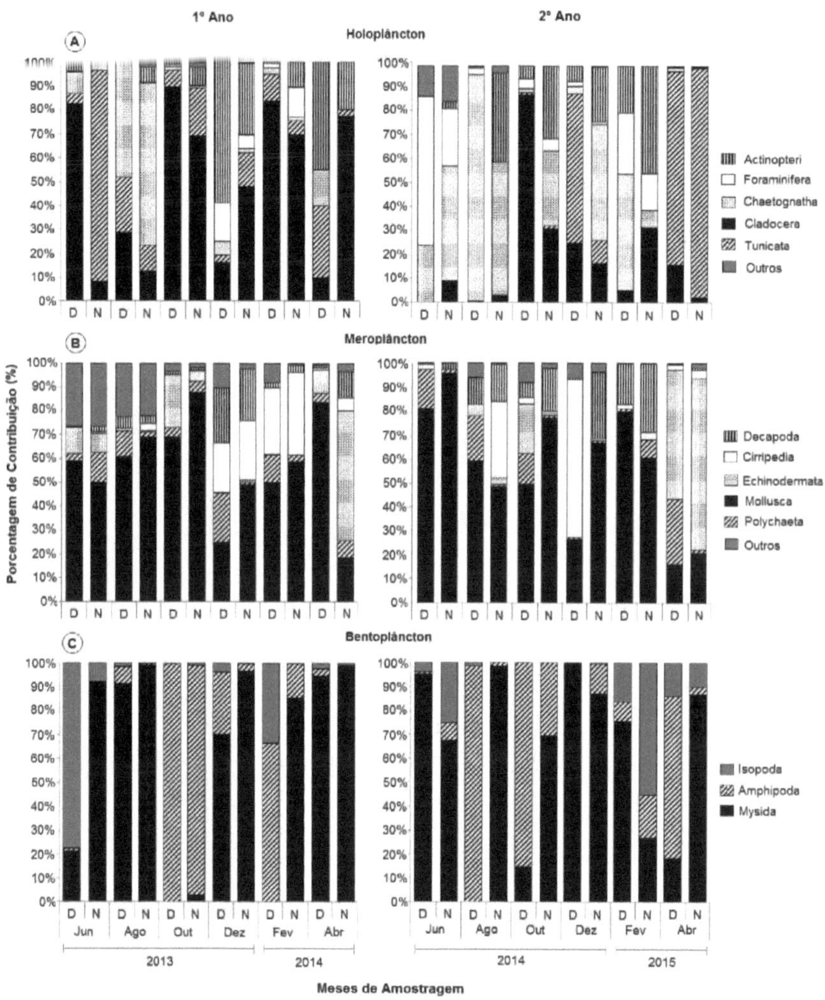

Figura 15: Percentage contribution of the main holoplankton, meroplankton and bentoplankton groups captured monthly during the two years of sampling in the Praia da Aviaçâo surf zone.

The proportions of the percentage contribution of the taxonomic groups in relation to the day and night periods are shown in Figure 16. Of the 27 taxonomic groups identified, 15 had a higher percentage of contribution during the day and 12 during the night. The holoplankton fauna showed an equivalent pattern in relation to the sampling period, with half contributing more during the day and half at night, however, the Tintinnina and Luciferidae groups occurred exclusively during the night period and juvenile Actinopteri reached 94% contribution also during the night. Among the meroplankton groups, the percentage was fairly homogeneous in relation to the sampling period, with seven contributing more during the day and five at night, with the Polychaeta (day) and Thalassinidae (night) standing out. As for the bentoplankton fauna, there were two groups with a higher percentage during the day and one at night corresponding to the Mysida group.

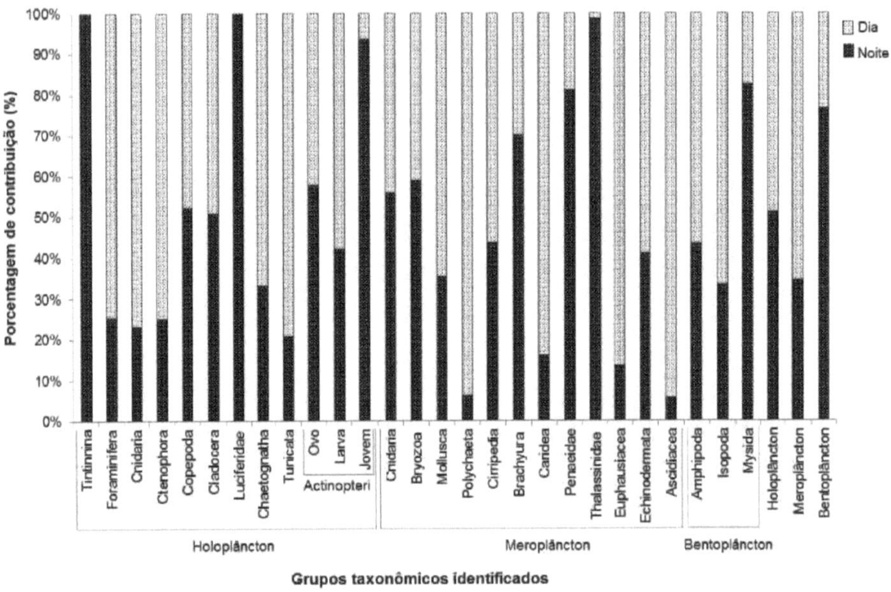

Figura 16: Percentage contribution of the main holoplankton, meroplankton and bentoplankton groups in relation to day and night sampling over two years of analysis in the Praia da Aviaçao surf zone.

6.2.4. Richness, Diversity and Equitability

The zooplankton fauna recorded in the surf zone throughout the study showed a minimum richness of 19 and a maximum richness of 32 during the day. At night, the values varied between 21 and 41.

The Shannon index in the first sampling year ranged from 1.51 to 3.48 (2.75± 0.75 bits.ind-^{1}) during the daytime period and from 0.89 to 3.05 (2.43± 0.8 bits.ind-^{1}) at night, while in the second year the values during the day ranged from 2.03 to 3.33 (2.75 ± 0.64 bits.ind-1) and at night from 1.69 to 3.43 (2.79 ± 0.64 bits.ind-i). Over the two years of analysis, the index reached higher values in the daytime collections, although the difference in diversity between the periods was low (Figure 17).

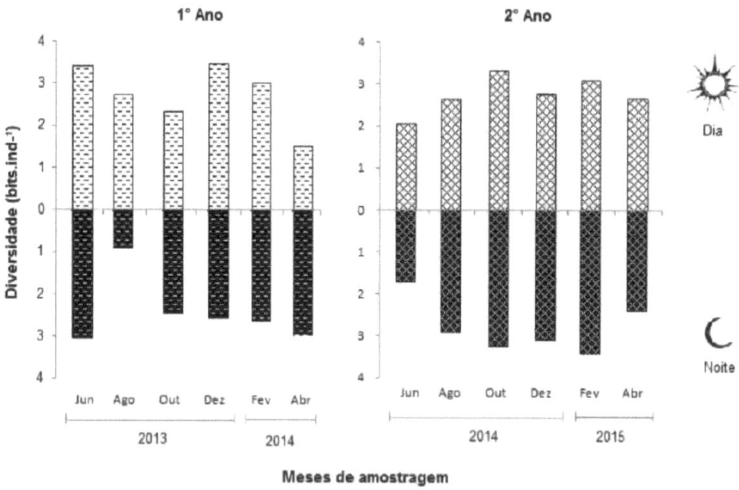

Figure 17: Shannon diversity index of the zooplankton fauna during the day and night periods of each sampling month over two years of analysis.

The evenness of the zooplankton fauna analyzed had a low amplitude during this study. For the first sampling year, it varied during the day from 0.24 (April/2014) to 0.55 (December/2013) and at night from 0.14 (August/2013) to 0.38 (October/2013). In the second year, the variation during the day was from 0.32 (June/2014) to 0.53 (October/2014) and at night from 0.27 (June/2014) to 0.55 (February/2015). Overall, during this study, the average equitability during the day was 0.44 (± 0.09) and 0.41 (± 0.11) for the night period.

6.2.5. Fauna Variation

The total abundance of zooplankton caught over the entire sampling period amounted to 143981.3 org.m^{-3} (5999.2± 13758). The holoplankton fauna was 12 times more abundant than the meroplankton fauna and 51 times more abundant than the bentoplankton fauna. The abundance of most of the groups recorded was lower during the day (49837.4 org.m-3) than at night (94143.9 org.m-3), with the fauna being almost twice as abundant at night as during the day (Table 3). During the day, the abundance reached 43823.7 org.m-3 (3652 ± 5240.4), 5473 org.m-3 (456.1 ± 1005) and 360.4 org.m-3 (30 ± 30.6) for the holo, grouper and bentoplankton fauna, respectively. During the night, the abundance values were 86515.1 org.m-3 (7209.6 ± 17797.1), 5366.5 org.m-3 (447.2 ± 991.6) and 2209.2 org.m-3 (184.1 ± 297) for the holo, mero and bentoplankton organisms, respectively.

The abundance records per month and per sampling period, together with an indication of the stage of development of the taxa identified, are presented in Annex 2 .

Table 3: Record of the total abundance values of the main zooplankton groups sampled over the study period. Clas. = Classification; D = Day; N = Night.

Clas.	Fee	Abundance of zooplankton fauna						
		1*	Year	2= Year		Sample Period		
		D	N	D	N	D	N	Total
Holoplankton		12.301,5	15.900,9	31.522,3	70.614,2	43.823,7	86.515,1	130.338,8
	Tintinnin	O,O	8,2	o,o	0,2	o,o	8,4	8,4
	Foraminifera	18,3	45,1	75,6	12,5	93,9	58,6	152,5
	Cnidaria	24,0	10,3	4,2	5,6	28,2	15,9	44,1
	Ctenophora	0,0	o,o	8,8	5,5	3,8	5,5	14,3
	Copepoda	11050.3	14.672,0	28.988,9	66.842,2	40.039,2	81.514,2	121.553,4
	Cladocera	116.2	439.2	1.926,6	3.529,0	2.042,9	3.968,2	6.011,0
	Luciferidae	0,0	o,o	0,0	0,5	0,0	0,5	0,5
	Chaetognatha	40,5	31,3	59,4	61,1	99,8	92,4	192,2
	Tunicata	1.034,0	600,6	427,6	106,6	1.461,6	707,2	2.168,9
Actinopteri	Egg	16,2	78,6	27,6	33,4	43,7	112,1	155,8
	Maggot	1,2	5,4	3,3	0,3	4,5	6,2	10,7
	Youth	0,7	9,2	0,3	16,7	1,0	25,9	26,9
Meroplankton		736,9	1.410,1	4.736,1	3.956,5	5.473,0	5.366,5	10.839,6
	Cnidaria	37,3	52,8	9,5	58,6	46,8	111,4	158,3
	Bryozoa	11,3	37,6	5,5	7,7	16,8	45,3	62,1
	Mollusca	308,0	828,4	1.492,3	1.015,5	1.800,3	1.843,9	3.644,2
	Polychaeta	89,9	91,1	1.117,7	53,2	1.207,5	144,3	1.351,9
	Cirripedia	97,3	176,5	128,2	151,0	225,5	327,5	553,0
	Brachyura	7,5	33,6	3,7	37,5	16,2	71,1	87,3
	Caridea	0,4	2,7	31,6	8,6	32,0	11,3	43,3
	Penaeidae	1,8	11,4	1,2	12,6	3,0	23,9	26,9
	Thalassinidae	o,o	36,2	0,7	48,5	0,7	84,7	85,4
	Euphausiacea	39,3	7,1	2,4	4,8	41,7	11,9	53,6
	Echinodermata	123,7	130,2	1.938,4	2.558,7	2.062,0	2.688,9	4.750,9
	Ascidiacea	20,4	2,2	0,0	0,0	20,4	2,2	22,6
Benthoplancton		171,0	1.689,9	189,3	519,3	360,4	2.209,2	2.569,5
	Amphipoda	10,4	47,9	37,4	20,9	47,8	68,8	116,6
	Isopoda	56,1	17,3	18,9	53,1	75,0	70,4	145,4
	Mysida	104,4	1.624,7	133,1	445,3	237,5	2.070,0	2.307,5
Invertebrate egg		126,1	47,6	54,1	5,5	180,3	53,1	233,4
Total		1.3335,5	19.048,5	36.501,9	75.095,4	49.837,4	94.143,9	143.981,3

With regard to the monthly abundance of zooplankton recorded in the first year, during the daytime period catches ranged from 528 org.m-^{3} (December/2013) to 5561 org.m-3 (April/2014) and at night from 1686 org.m-3 (April/2014) to 9105 org.m-3 (February/2014). Collections during the day totaled 1,335 org.m-3 and at night 1,948 org.m-3 (Figure 18a).

In the second year analyzed, the abundance of daytime fauna ranged from 384 org.m-3 (August/2014) to 17243 org.m-3 (February/2015) and at night from 504 org.m-3 (June/2014) to 67269 org.m-3 (April/2015). In this second year, the total fauna caught during the day was 36502 org.m-3 and at night 75095 org.m-3 (Figure 18b).

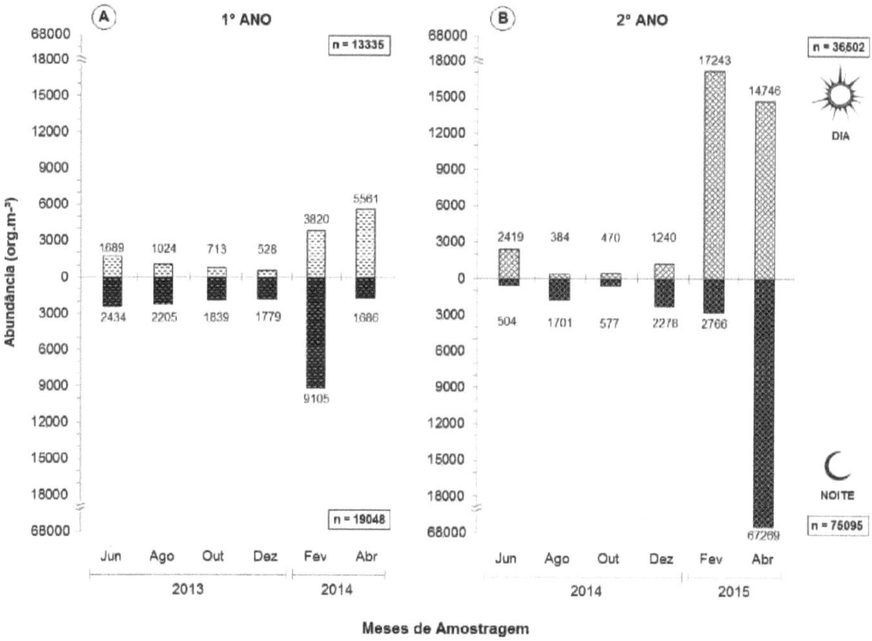

Figure 18: Monthly abundance data (org.m-3) of the zooplankton fauna caught during the day and night periods in the Praia da Aviaçao surf zone.

Analyzing the average monthly abundance throughout the study, the lowest catch of the zooplankton fauna occurred in October, both during the day (592 org.m-^{3} ± 172) and night (1208 org.m-3 ± 892), while the highest catch was recorded in February (10531 org.m-3 ± 9491) and April (34478 org.m-3 ± 46374) during the day and night, respectively (Figure 19).

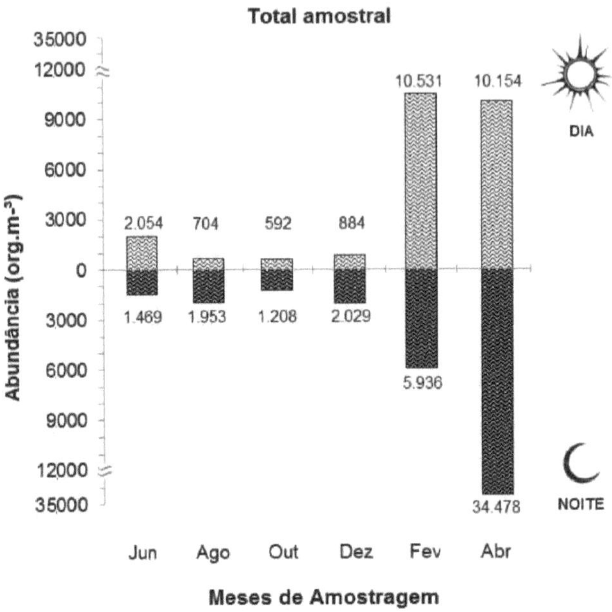

Figure 19: Average monthly abundance data (org.m-3) of the zooplankton fauna caught during the day and night periods in the Praia da Aviaçâo surf zone over two sampling years.

In all the months analyzed, the holoplankton fauna was more abundant than the meroplankton fauna and the bentoplankton fauna. The monthly abundance of holoplankton throughout the study varied during the day from 329 org.m-3 (December/13) to 16980 org.m-3 (February/15) and during the night from 438.6 org.m-3 (June/14) to 63285.8 org.m-3 (April/15) (Figure 20a, 20b).

The meroplankton community, although not shown graphically (Figure 20a, 20b), also showed variations in abundance throughout the study, both during the daytime period between 10.3 org.m-^{3} (August/14) and 3557.4 org.m-^{3} (April/15), and at night with catches between 40.1 org.m-3 (October/14) and 3570 org.m-3 (April/15).

With regard to the abundance of benthoplanktonic organisms in the water column, there were also variations over the months analyzed, with daytime abundances ranging from 0.1 org.m-3 (December/14) to 82.3 org.m-3 (June/14) and nighttime abundances ranging from 3.2 org.m-3 (October/14) to 980.2 org.m-3 (August/13) (Figure 20a, 20b).

In general, in most months the abundance of the holo, mero and bentoplankton fauna recorded in the upwelling zone did not exceed 4000 org.m-3. However, there were variations over the months, especially in February and April of both sampling years (Figure 20a, 20b). This dispersion in values was noted mainly for the holoplankton organisms caught in the second sampling year during the nocturnal period, but smaller-scale variations were also observed for the meroplankton fauna.

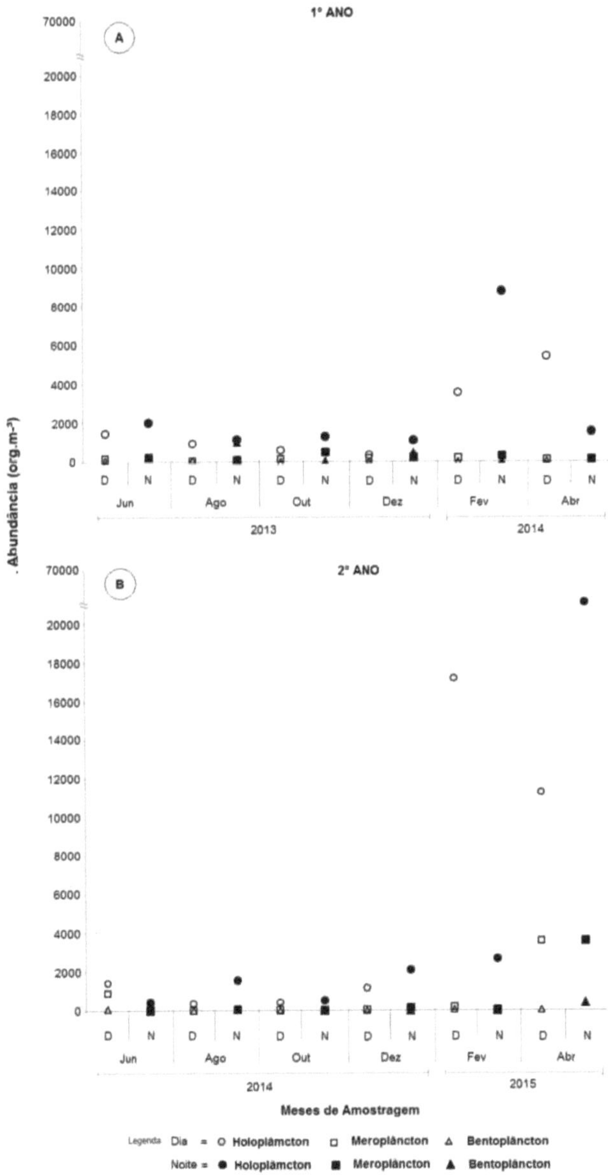

Figure 20: Monthly abundance data (org.m-^{3}) of the holoplankton, meroplankton and bentoplankton fauna captured during the day and night periods in the surf zone of Praia da Aviaçâo over two years of sampling.

When analyzing the variation in the abundance of the zooplankton fauna over the two years of the study, in relation to life habits, it can be seen that for most of the year the average abundance of holoplankton organisms did not exceed 2000 org.m-3 at both times of the day, but in February and April the abundance exceeded 4000 org.m-3, reaching a maximum of 32412 org.m-3 at night. The meroplankton fauna showed a similar pattern, but with much lower values, not exceeding 600 org.m-3 in all months, with the exception of April when it reached 1825.7 org.m-3. The benthoplankton showed a constant variation in abundance over all the months and periods analyzed. In general, the fauna did not exceed 300 org.m-3, with the highest abundance of 520.3 org.m-3 in August during the night period (Figure 21).

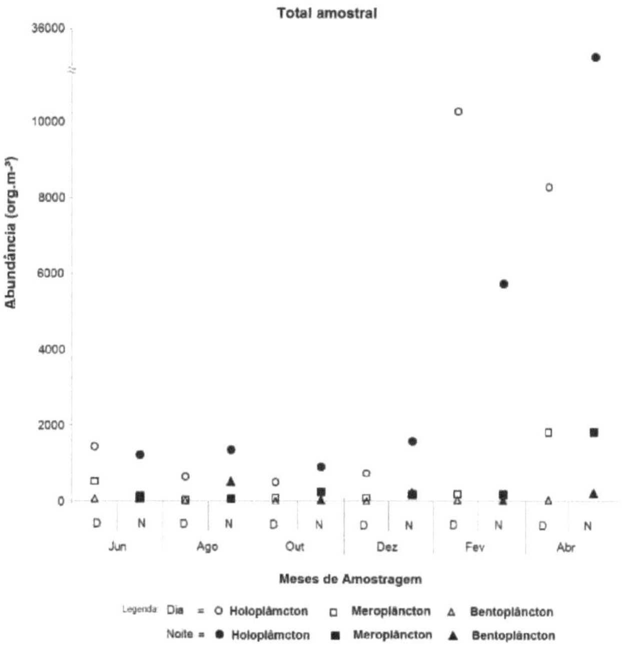

Figure 21: Average monthly abundance data (org.m-3) of the holoplankton, meroplankton and bentoplankton fauna captured during the day and night periods in the Praia da Aviaçâo surf zone over two years of sampling.

In general, the average abundance of the zooplankton fauna captured throughout the study was 5999.2 org.m$^{(-3)}$ (± 13758), with 4153.1 org.m$^{(-3}$± 5766) and 7845.3 org.m$^{(-3}$± 18844) being recorded during the day and night, respectively. As for the multi-year comparison, the abundance values in the daytime period in the second sampling year were 2.7 times higher than in the first year, ranging from 6084 to 2223 org.m-3, while in the nighttime period the variation recorded was 3.9 times higher, with 12516 and 3175 org.m-3, respectively (Figure 22a, 22b).

As mentioned above, in both sampling periods the Copepoda taxon was found to be dominant in the holoplankton fauna, as well as in the entire zooplankton community, becoming even more important numerically in the second sampling year. Other components that stood out in the holoplankton were tunicates and cladocerans. In the first sampling year, tunicates were more expressive with 172.3 org.m-3 (day) and 100.1 org.m-3 (night) and, in the second year, cladocerans were caught in greater abundance with 321.1 org.m-3 (day) and 588.2 org.m-3 (night). In the meroplankton, the dominant taxa in average abundance were molluscs, polychaetes and echinoderms, with the annual variation in echinoderms being more evident when compared to the other meroplankton groups. Among the bentoplankton, the greatest annual variation occurred in misidaceans, with greater abundance in the first year, during the nocturnal period, when compared to the second sampling year (Figure 22a, 22b).

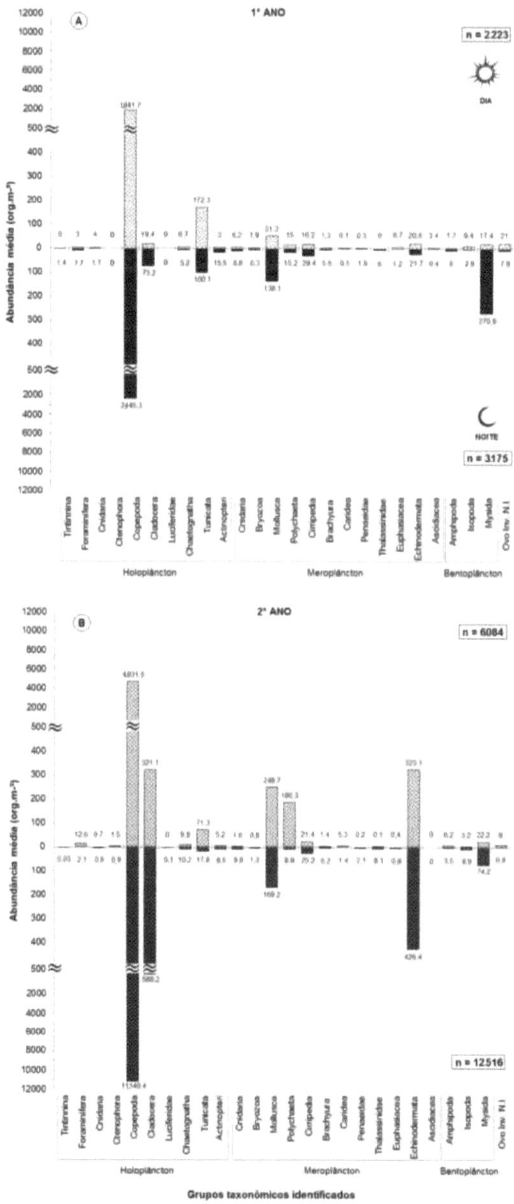

Figure 22: Average abundance data (org.m-^{3}) of the taxonomic groups of the holoplankton, meroplankton and bentoplankton fauna captured during the day and night periods in the surf zone of Praia da Aviaçâo during the first and second sampling year.

In general, when analyzing the average abundance of the zooplankton fauna throughout the sampling period, the dominance of certain holoplankton groups (Copepoda, Cladocera and Tunicata), meroplankton (Mollusca, Polychaeta and Echinodermata) and bentoplankton (Mysida) is clear. There was also a large variation in the overall abundance of the zooplankton community between the collection periods (daytime vs. nighttime), with the nocturnal fauna being about twice as large as the daytime one, with 7845 and 4154 org.m-^{3}, respectively (Figure 23).

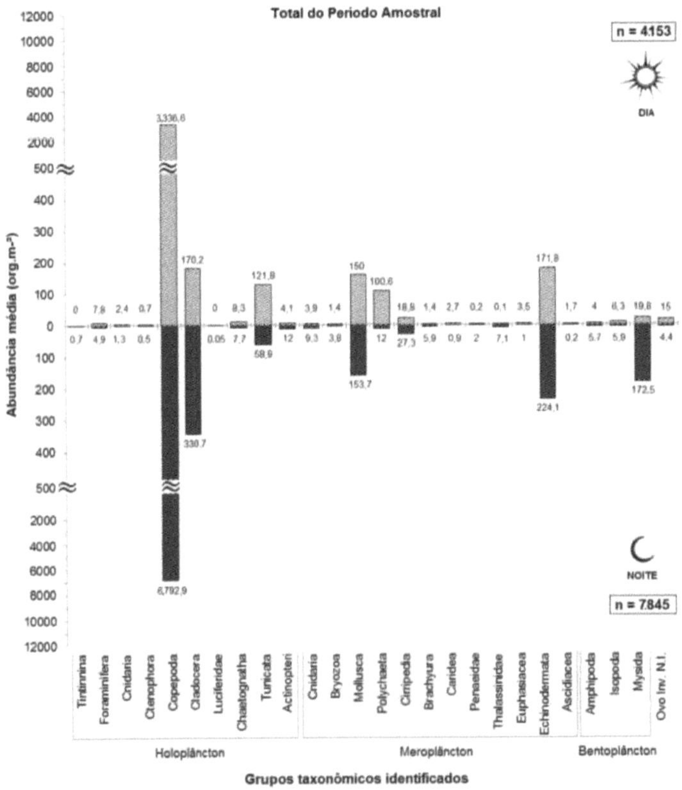

Figure 23: Average abundance data (org.m-3) of the taxonomic groups of the holoplankton, meroplankton and bentoplankton fauna caught during the day and night periods in the Praia da Aviaçâo surf zone over two years of sampling.

6.2.5.1. Statistical Analysis

GLM (*Generalized Linear Model*) analyses relating the biotic and environmental variables to the total zooplankton abundance data were used to verify the effect of the factors on the abundance of the fauna (Table 4). The variables Community (C) and Month of sampling (M) were significant ($p < 0.001$) in both analyses, both when verified individually and in the final model as a whole, showing that zooplankton abundance in the upwelling zone shows significant differences both in relation to community composition and the months of sampling. However, the factors Sampling period (P) and Water temperature (T) were not significant enough to have influenced local zooplankton abundance, and these factors were removed from the final model (Table 4 - Analysis 1). Thus, the final model obtained was C+M (Table 4 - Analysis 2).

Table 4: *GLM* (*Generalized Linear Model*) analyses applied between biotic and environmental variables in relation to zooplankton abundance data. Analysis 1 refers to the individual analysis of each variable to select the model, while Analysis 2 applies the statistics to the model as a whole. (R^2 = Coefficient of determination; (p) Significance of the analysis; C = Community; P = Sampling period; M = Sampling month; T = Water temperature; * indicates significance; - indicates variable removed from the model).

Zooplâncton Total			
Análise 1			
	R^2	**R^2** (Ajustado)	**p**
Modelo	0,288	0,235	<0,001*
C			<0,001*
P			0,123
M			<0,001*
T			0,299
			Modelo selecionado: C-M
Análise 2			
	R^2	**R^2** (Ajustado)	**p**
Modelo	0,573	0,490	<0,001*
C			<0,001*
P			----
M			<0,001*
T			----
C*M			<0,001*

6.2.5.2. Multivariate Analysis

The *NMDS* (*Non Multi Dimensional Scaling)* multivariate analysis produced a two-dimensional image illustrating the similarities, in terms of zooplankton abundance, between the samples related according to the climatic seasons. The groups were formed by *cluster* analysis and plotted on the *NMDS* (green lines). To apply the season factor, it was necessary to average the zooplankton abundance recorded in both periods (day and night), as well as the average between the October and December/2013 campaigns corresponding to spring/2013, April and June/2014 corresponding to fall/2014 and October and December/2014 corresponding to spring/2014. The other samplings such as June/2013 correspond to fall/2013, August 2013/2014 to winter of the same years, February 2014/2015 to the summer season of the respective years and, finally, April/2015 to the fall of the same year.

Thus, the *NMDS* related to the seasons made it possible to observe the distribution of the samples in a few groups: (1) fall, winter and spring of 2013/2014; (2) summer of 2014/2015 and; (3) fall of 2015 alone. In this analysis it was also possible to see that the abundance of the zooplankton fauna showed a similarity of 80% between the climatic seasons that make up each group formed, with a similarity of > 60% between groups (1) and (2) and < 40% between group (3) and the others (Figure 24).

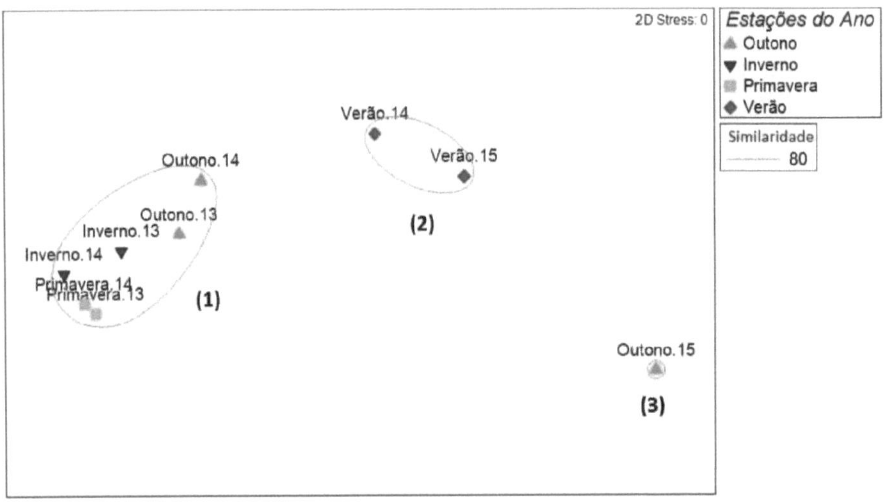

Figure 24: Distinction of zooplankton samples by *NMDS* analysis. The samples were represented by the season factor with four categories: fall (2013, 2014 and 2015), winter (2013 and 2014), spring (2013 and 2014) and summer (2014 and 2015). Groups (1), (2) and (3) were formed by 80% similarity.

7. Discussion

7.1. Environmental variables

The environmental variables recorded in the zone analyzed showed variations over the course of this study. With regard to the water temperature variable, it showed a pattern of seasonal variation, i.e. maximum values during the summer and minimum values during the winter, as well as oscillations throughout the seasons marked by a progressive increase during the spring and a decline in the fall. This pattern of variation, typical of subtropical zones, was also observed by Fontes (2010) in the central region of Baixada Santista, where the water temperature reflected the local atmospheric conditions.

According to Buthelezi (2012), the salinity content of seawater in upwelling zones is also characterized by constant fluctuations, thus corroborating the data obtained in this study. According to this author, this is due to the fact that the upwelling zone occupies a transition area between the terrestrial and marine environments, where in many cases continental watercourses flow. Therefore, the low salinity levels found in the study area may indicate greater intrusion of freshwater into the system by river and rainwater runoff, since the sampling point was influenced by two streams that drain continental water from part of the municipality. The high salinity values can be attributed to the high temperatures and, consequently, high evaporation characteristic of the summer and, occasionally, the other seasons.

7.2. Zooplankton community

7.2.1. Richness, Diversity and Equitability

The richness of taxonomic groups observed in the surf zone analyzed was high, comprising 116 taxa, when compared to the southern region of the country, where the number of groups recorded by Avila *et al.* (2009), at Tramandai Beach/RS, was 1.4 times lower, as well as at Pontal do Sul Beach/PR where Straub *et al.* (2014) captured 5.3 times fewer taxonomic groups in relation to those recorded at present. The greater number of taxa presented in this study may be related to the analysis of richness which also included nocturnal fauna and benthic organisms in their adult phase, which were not addressed by Avila *et al.* (2009) and Straub *et al.* (2014). In addition, there are differences in environmental conditions between the areas due to the latitudes at which they are located, where Praia da Aviaçao (24°S) has more tropical characteristics compared to Praia de Tramandai (30°S) and Praia de Pontal do Sul/PR (25°S), which consequently alter the composition of the local zooplankton fauna. However, despite the environmental contrasts inherent to latitude, Costa *et al.* (2011) observed a lower richness at Ajuruteua Beach/PA (01°S) than at Praia da Aviaçao/SP, where the record was 2.4 times lower. In addition, a lower value was also observed off the coast of Pernambuco, where Melo *et al.* (2010) captured 1.3 times fewer taxonomic groups. In two locations on the African coast, Buthelezi (2012) caught similar numbers of taxonomic groups during the day and night periods as in the present study, 132 and 108 taxa in the St. Lucia and Mhlathuze surf zones, respectively, and in Oregon, on the west coast of the United States, Marin Jarrin & Shanks (2010) recorded a total of 105 species during both periods (day and night). However, there are certain difficulties in establishing comparisons of richness due to different methodologies and sampling efforts.

With regard to the richness of the fauna between the different times of the day, this study recorded on average higher values at night than during the day, suggesting that certain planktonic organisms have moved to the surface and that, therefore, the occurrence of the zooplanktonic fauna in the surf zone is different during the day and night periods. Marin Jarrin & Shanks (2010) also pointed out that the zooplankton fauna was richer during the night. This differentiation is associated with endogenous factors (reproduction and physiological aspects) (Lasiak & Mclachlan, 1987) or exogenous factors (light, temperature, salinity, predation, differential abundance between taxa, etc.) (Clutter, 1967; McLachlan, 1983; Takahashi & Kawaguchi, 1997; Heard *et al.*, 2006).

In general, the zooplankton community of the Praia da Aviaçao surf zone shows little variation in diversity, following a seasonal pattern, especially when analyzing the second sampling year. The diversity and richness of the fauna decreases slightly from the summer onwards, followed by an increase in the spring; however, this pattern was not evident in the first year. Similar data was recorded by Avila *et al.* (2009), Marin Jarrin & Shanks (2010) and Buthelezi (2012) where this pattern was equivalent for the southern Brazilian region, the American west coast and Africa, respectively. In this study, the maximum diversity of fauna, with the exception of copepods, obtained during the daytime period was 1.3 times greater than that recorded by Avila *et al.* (2009) in the southern region, being 3.48 bits.ind-^{1} and 2.6 bits.ind-$^{(1)}$, respectively. However, despite the differences between the results, both studies found maximum diversity during the summer season.

In relation to the day and night periods, the diversity of the fauna remained practically stable, with the exception of August/2013, reflecting a relatively constant community. This low difference in diversity between the periods may be related to the frequent capture of misidaceans (spots) (Clutter, 1967; McLachlan & Brown, 2006) during the night and which, consequently, may have reduced the diversity index due to the numerical dominance of the group during the night period. The same pattern was observed in southern Africa by Buthelezi (2012), where the author recorded diurnal and nocturnal diversity values similar to those of the present study, where, on average, the index during the dark period was lower than during the light period, also due to the incursion of bentoplankton organisms into the water column. On the Oregon coast, Marin Jarrin & Shanks (2010) also obtained similar results, where the zooplankton fauna during the day was significantly more diverse than at night.

However, more in-depth comparisons cannot be made due to the difference in the taxonomic level of identification, method and sampling effort used in each survey, as well as the particularities of the environmental conditions in the surroundings of the different regions.

In the first year of sampling, equitability was higher during the day compared to the night, with the exception of October/2013 and April/2014, showing greater homogeneity during the day. This difference may be related to the greater density of organisms present in the plankton during the night, which consequently reduces the uniformity of the zooplankton fauna. In the second year, equitability in daytime sampling was highest in June/14, October/14 and April/15, while at night the greatest homogeneity corresponded to the months of August/14, December/14 and February/15. In relation to the two years of sampling, the zooplankton fauna in the second year showed greater numerical homogeneity between species than in the first year.

7.2.2. Temporal Variation

The zooplankton community recorded in the surf zone of Praia da Aviaçao showed significant temporal responses, where the variation in terms of abundance over the sampling months was proven using the generalized linear model (*GLM)* employed in this study. This variability was also partly demonstrated in the multivariate analysis plotted in *NMDS,* in an attempt to visualize seasonal patterns in the community, where the samples corresponding to the summer months were similar (80%), as were the other seasons (spring, autumn and winter).

According to Nielsen & Munk (1998), planktonic fauna reflects seasonality due to changes in environmental factors, mainly related to water temperature, which is responsible, for example, for changes in food availability. Considering this statement, it is possible to partly justify the dissimilarity recorded between summer and the other seasons, given that in summer the water in the surf zone was on average 6.5°C warmer than in the other seasons.

However, it was not statistically significantly associated with the environmental variable temperature in terms of abundance. In this way, answers that justify the temporal variation of the planktonic fauna in the area studied are difficult to interpret due to the analysis of isolated variables, and interpretation in conjunction with various environmental and ecological descriptors is more appropriate, for example, climatic and hydrodynamic factors, chemical composition of the water, dissolved oxygen and nutrient availability. However, these parameters were not included in the sampling design of this study, thus limiting the interpretation of temporal variations in the zooplankton community.

Using multivariate statistical techniques, it was also possible to see the isolation of the sample corresponding to autumn 2015 (April 2015), where it showed low similarity (< 40%) in relation to the other seasons, and this differentiation can be explained by the highest peak in abundance obtained throughout the study. This high abundance probably indicates the capture of clusters of organisms forming *patches*. This type of inhomogeneous distribution is common in the zooplankton community of upwelling zones and can change daily or over the course of hours, causing very significant effects on sampling (McLachlan & Brown, 2006; Vilano, 2016 verbal information). The densification of animals can be promoted by the action of waves on the coast, which can move planktonic fauna from other locations to the internal upwelling zone, consequently trapping organisms for a certain period of time in the region. Associated with this, the displacement and densification of the fauna can also be caused by the action of return or longitudinal currents to the coast (McLachlan & Hesp, 1984; Buthelezi, 2012; Vilano, 2016 verbal information) which, according to Souza (1997), are frequent in Praia Grande/SP. The combination of these factors may be related to the constant fluctuations in the abundance of the zooplankton fauna, but according to Vega-Pérez (1993) biological processes such as predatory pressure can also influence the abundance and distribution of organisms.

During the sampling period, the holo, mero and bentoplankton fauna that make up the zooplankton community were significantly different in terms of abundance, and this differentiation was confirmed by the *GLM*. The same pattern is described in similar studies carried out in different surf zones along the Brazilian coast (Bersano, 1994; Avila *et al.*, 2009; Melo *et al.*, 2010; Costa *et al.*, 2014a), as well as in other regions of the world such as the North Atlantic (Stull *et al.*, 2016) and the Pacific Ocean (DeLancey, 1987; Marin Jarrin & Shanks, 2010). This difference is probably due to the presence of the Copepoda group, which outnumbered the meroplankton fauna in both sampling periods; however, if the copepods were disregarded from the analysis, the meroplankton fauna would be pre-eminent. In general, the dominance of this group is reported by several authors, such as Bjornberg (1981), Huys & Boxshall (1991), Vega-Pérez (1993), Boltovskoy (1999), Bradford-Grieve *et al.* (1999), Cavalcanti & Larrazàbal (2004), Sant'Anna & Bjornberg (2006), Melo *et al.* (2010), Pereira (2010) and Martinelli Filho (2013), making it the best known group of marine zooplankton in the South Atlantic.

In general, in turbulent zones, the pattern of high abundance of the Copepoda group is maintained due to the high flexibility and ability to adapt to environmental variations that these organisms have (Huys & Boxshall, 1991). Some studies such as those by Brown & Mclachlan (1990), Mclachlan & Romer (1990), Franks (2001) and Freund *et al.* (2006) show that the group responds reproductively to the high availability of nutrients and phytoplankton biomass in the region, positively influencing the abundance of these organisms as described here and in other locations. In addition to this, Pinheiro *et al.* (2011) also relate the strong local hydrodynamism responsible for the suspension of typically benthopelagic species and further increasing the abundance of the group in the superficial layers of the water column. However, despite the remarkable representativeness of copepods, it is believed that the dominance of zooplankton groups in the surf zones can vary depending on the characteristics and environmental conditions of the surroundings, with Mysidacea and Brachyura fauna outnumbering Copepoda abundance in certain places, for example in the research carried out at Praia do Cassino/RS by Bersano (1994), in South Africa by Buthelezi (2012) and in Ecuador by Marin Jarrin *et al.* (2016), but this pattern was not observed in the present study.

The zooplankton fauna of the surf zone recorded during this study exceeded, in terms of abundance, around 1.2 times the concentration found in deeper coastal zones (from 18 to 35m), such as those recorded off the coast of Ubatuba (Vega-Pérez, 1993; Martinelli Filho, 2013) and Santos Bay (Pereira, 2010). This observation was also presented by Stull *et al.* (2016) when they analyzed and compared the planktonic fauna of beaches in North Carolina, United States, with adjacent regions of greater depth.

According to Stull *et al.* (2016), the upwelling zones also have on average twice as much phytoplankton biomass as the shelf waters. Therefore, the high concentration of planktonic fauna can be explained by the complex characteristics intrinsic to shallow turbulent zones, such as the strong hydrodynamics established by breaking waves and currents, which promote the constant redistribution and circulation of nutrients retained in the sediment into the water column, thus favoring phytoplankton development and, consequently, the availability of food for zooplanktonic fauna (McLachlan, 1980; Stull *et al.*, 2016). In addition, the high energy favors the semi-permanent accumulation of diatoms in the surf zones, this type of formation being specific to exposed and dissipative beaches (Campbell & Bate, 1991a; 1991b; 1997; Defeo *et al.*, 2009), such as Praia Grande/SP. In this way, the frequent concentration of diatoms increases the phytoplankton biomass, providing even more food for primary consumers, positively reflecting on zooplankton abundance (Gianuca, 1983; Brown & McLachlan, 1990; Bersano, 1994; Rorig *et al.,* 1997; Hewson *et al.,* 2001; Kahn & Cahoon, 2012; Stull *et al.*, 2016). In this way, it is suggested that the zooplankton fauna of the zone of upwelling analyzed, due to its high abundance, represents an important food resource for planktivorous organisms in this region, corroborating the study by DeLancey (1987).

At the same time, the region's high rainfall (Alvares *et al.*, 2013) favors the greater intrusion of fluvial water into the analyzed breakwater zone, through artificial channels that drain towards the sea (Souza & Cunha, 2009; 2012). This fact promotes the transport of nutrients and organic matter of continental origin into the system, where they are trapped by the action of the waves and made available for the growth of the algal community (Kahn & Cahoon, 2012), influencing the development of the planktonic fauna (Costa *et al.*, 2011). It is also worth mentioning that the study area may eventually receive nutrients from the Santos/Sào Vicente estuarine system, because according to Souza (1997), depending on the local hydrodynamic conditions, there are plumes coming from Santos Bay which, when they go around Ponta de Itaipu, reach the Praia da Aviaçà coastline, making even more nutrients available for phytoplankton growth and the entire local trophic chain.

Among the holoplankton groups, tintinids were scarce in the study area, having been caught in low abundances only at night during the spring and summer. However, Costa *et al.* (2014a) caught tintinids very frequently during the day, in the surf zone of Ajuruteua Beach/PA, but the net used had a mesh size of 64μm. Therefore, the methodology used in this study (180μm mesh) may have underestimated the real occurrence of this group in the surf zone. It may be necessary to use a finer mesh net in order to capture smaller organisms more efficiently. This aspect was demonstrated by Martinelli Filho (2013), in a study carried out in Flamengo Bay (Ubatuba/SP), where he recorded an abundance of tintinids seven times greater when using 20μm mesh nets.

The Foraminifera group did not have a high percentage of contributions among the holoplankton captured, but throughout the study the group was frequently captured during both sampling periods (day and night). The high frequency of occurrence was also recorded by Costa *et al.* (2014a) during the day on the Pará coast. The foraminifera of the genus *Globigerinoides* occurred in the Praia da Aviaçao breakwater zone in all seasons, except in the fall and winter of the first sampling year; the other genera (*Globigerina* and *Orbulina*), as well as unidentified individuals, occurred sporadically. In general, the abundance of the group was low throughout the study.

The groups of holoplanktonic cnidarians were present in the surf zone sporadically, with Diphyidae being absent in winter, while the other families were only caught in this season during the first year of sampling.

Cladocerans were present in both sampling periods throughout the study, but their occurrence was not constant: in the first year they were caught very frequently, while in the second year their presence in the surf zone was sporadic. The abundance of the group was also not stable throughout the year, with peaks unrelated to the climatic seasons. Despite this, studies have shown a positive relationship with temperature, where higher abundances of the group reflected seasonal variation, as described by Miyashita *et al.* (2010; 2011) and Martinelli Filho (2013). According to Atienza *et al.* (2007) and Martinelli Filho (2013), cladocerans show a rapid reproductive response as a result of their efficient reproductive strategy (pedogenesis) and high growth rate. Furthermore, in unfavorable environmental conditions, cladocerans are able to change their reproductive mode and produce resistance eggs that remain in the sediment until conditions are ideal for hatching (Atienza *et al.*, 2007). This combination of factors possibly explains the inconstant occurrence of cladocerans, with high peaks of abundance and absence of the group at certain times of the year in the Praia da Aviaçao breakwater zone.

Copepods were caught throughout the study in the Praia da Aviaçao breakwater zone, but were more abundant at night and during the summer and autumn seasons. In this study, they were considered the dominant group among the zooplankton fauna analyzed, accounting for 84% of the total abundance. A similar result was obtained by Costa *et al.* (2014a) at Praia de Ajuruteua/PA, where copepods accounted for 83% of the planktonic organisms; however, Costa *et al.* (2011) previously recorded a 43% contribution from the Copepoda group in relation to the zooplankton fauna on the same beach on the Para coast, thus showing that despite the group's high contribution, its oscillation in turbulent zones is common.

The order Calanoida was more abundant when compared to Cyclopoida, Harpacticoida and Poecilostomatoida, and was also described in research on marine zooplankton by Bjornberg (1981), DeLancey (1987), Bradford-Grieve *et al.* (1999), Cavalcanti & Larrazàbal (2004), Avila *et al.* (2009), Pereira (2010) and Stull *et al.* (2016), as the order with the highest species diversity among the Copepoda group. Specifically in upwelling zones, the high representativeness in the plankton of the Copepoda group has been described off the coast of South Africa (Buthelezi, 2012), Portugal (Frazao, 2015), the east and west coasts of the United States (DeLancey, 1987; Marin Jarrin & Shanks, 2010; Stull *et al.*, 2016). On the Brazilian coast, Avila *et al.* (2009) described the abundance and importance of the group at Tramandai Beach/RS, with the maximum abundance recorded by the author (1884org.m-$^{3)}$ being lower than the result obtained in this study (16,909.3 org.m-3/day and 59,646 org.m-3/night). Pinheiro *et al.* (2011) also found lower abundance (36380 org.m-3) at Praia de Ajuruteua/PA, while

Costa *et al.* (2014b) found much higher concentrations (330470 org.m-3) at the same location on the north coast. On Pontal do Sul Beach, Straub *et al.* (2014) also found a lower abundance (45753 org.m-3) than on the beach analyzed in this study. Despite the differences in the abundances listed for the Copepoda group, it is extremely complex to establish comparisons between the study sites due to the different methodologies used, as well as the sampling effort.

The Chaetognatha and Appendicularia groups were present in the study area throughout the year and sampling period and are therefore considered to be resident in this region. However, high concentrations were observed during the warmer months, such as summer for the ketognaths and spring and summer for the appendicularians. With regard to ketognaths, similar patterns of frequency of occurrence have been recorded in the offshore areas of Rio Grande do Sul and Parà by Avilla *et al.* (2009) and Costa *et al.* (2014a), respectively. This group in particular is considered to be one of the main representatives of the coastal environment, where its abundance is influenced by temperature and food availability (Liang & Veja-Pérez, 1994; Marazzo & Nogueira, 1996; Araùjo & Ribeiro, 2005) and these are therefore the possible reasons for the frequent capture of the *Sagitta* genus at Praia da Aviaçào.

Organisms from the Appendicularia group belonging to the *Oikopleura* genus, according to Esnal (1999), are also frequently related to coastal regions, with peaks in abundance being directly related to high temperature water (Campos, 2000; 2004), thus corroborating the data obtained in this study. In the region analyzed, equivalent data was obtained by Pereira (2010) where, in Santos Bay, he recorded a high frequency with maximum abundance in summer.

The tunicates of the order Doliolida occurred sporadically, and when they were present, the highest abundance recorded was during the summer in the daytime of the first sampling year. Similar results were observed by Avila *et al.* (2009) who found that the presence of these organisms was more evident during the day at this time of year. The salps of the genus *Thalia* did not show a clear pattern of occurrence, but their presence was more evident and constant during the winter season of the first sampling year, in both periods (day and night).

The presence of the Actinopteri group, in the form of eggs and larvae, was considered to be resident due to constant capture, thus indicating the importance of the area in the development of the ichthyoplankton fauna present in shallow waters, also described by Bennett (1989), Layman (2000), McLachlan & Brown (2006), Vasconcellos *et al.* (2007) and Stull *et al.* (2016). Most of the juveniles caught, such as the genera *Atherinella* and *Mugil*, are cited in the literature for completing their life cycle in the surf zones (Paiva Filho & Toscano, 1987; Monteiro-Neto *et al.*, 2003; Favero & Dias, 2015). In general, the Actinopteri group (including juveniles) was less evident during the winter months, with the greatest abundance recorded in the warm seasons, as also observed on beaches close to the study area (Paiva Filho & Toscano, 1987; Giannini & Paiva-Filho, 1995) and other regions of the Brazilian coast (Pessanha *et al.,* 1999; Barreiros *et al.,* 2004; Godefroid *et al.*, 2004; Vasconcellos *et al.,* 2007). According to Paiva-Filho & Toscano (1987) this pattern of occurrence in the region happens for juvenile Mugilidae, probably after spawning in the open sea and migrating to exposed beaches.

The variation in occurrence throughout the year may be related to physiological characteristics, such as reproductive cycles, hydrodynamic factors (responsible for transporting eggs and larvae from breeding areas to recruitment zones), the instability characteristic of turbulent zones (which makes it difficult for many organisms to remain there constantly), as well as changes related to other environmental variables (Ricklefs, 1996; Vazzoler, 1996; Layman, 2000; Mariani, 2001; Potter *et al*, 2001; Barreiros *et al.,* 2004; McLachlan & Brown, 2006; Favero & Dias, 2015).

Of the meroplankton groups, the cnidarians were practically constant in the Praia da Aviaçâo surf zone, with the Tubulariidae being caught exclusively in the winter of the first year. The larvae of bryozoans, cirripedia and echinoderms (echinopluteus) occurred in practically every month and are considered to be resident in the area. However, the larvae of echinoderms of the Asteroidea and Holothuroidea classes occurred migratory and sporadically (see Annex 2), respectively.

The constant capture of benthic invertebrate larvae, which colonize the rocky substrate as adults, is probably associated with the hydrodynamics of the surf zone (Mclachlan, 1980; MacMahan *et al.*, 2009; Shanks *et al.*, 2010). According to Souza (1997), the predominant longitudinal current at Praia da Aviaçao flows in a south-westerly (SW) direction, i.e. from Morro do Itaipù (the northern end of the shore) towards the city of Mongaguà (located to the south), and the circulation is therefore capable of transporting larvae released by organisms attached to the rocky shore (Morro do Itaipù) located 3.5km to the north (N) of the sampling point. In studies carried out in the state of Rio Grande do Sul, on the beaches of Cassino (Bersano, 1994) and Tramandai (Avila *et al.*, 2009), the constant presence of cirri was also recorded during the daytime period, although the possible origin of the larvae was not mentioned by the authors of these studies.

The larvae of gastropods, bivalves and polychaetes (Spionidae) were caught with high frequency, and are considered to be resident in this area. The catches of both groups were constant and did not vary much over the seasons, but the larvae of the other polychaete families occurred sporadically, with a greater variety of families in the spring season of the first sampling year. In all the catches, the Mollusca group also showed no variation in abundance in relation to the sampling period, however, polychaetes were 8.3 times more abundant during the day. According to Stull *et al.* (2016), the constant presence of meroplanktonic larvae from the groups mentioned above reflects the high availability of food offered by turbulent zones, which significantly and positively influences the continued spawning of benthic invertebrates.

In coastal environments, the frequent occurrence of Mollusca and Polychaeta larvae in the plankton is described in several studies (Vega-Pérez, 1993; Mello *et al.*, 2010; Martinelli Filho, 2013; Pereira, 2010) and specifically in the upwelling zone, they were also frequently recorded by Costa *et al.* (2011; 2014a) on the Para coast. Among the molluscs, throughout the study, Gastropoda larvae were predominant in relation to Bivalvia, and this relationship was also observed in Santos Bay by Pereira (2010).

Regarding polychaetes, the predominance of Spionidae larvae was also presented by

Pereira (2010) in Santos Bay, and the preponderance of this family in both areas can be justified by the similar characteristics of the sedimentary substrate, where specifically the beach analyzed has a high content of fine sand (Souza, 1997), a substrate mentioned as a favorite for many species included in the family (Fauchald & Jumars, 1979; Amaral & Nono, 1981). Jumars, 1979; Amaral & Nonato, 1981), thus justifying the constant occurrence of Spionidae larvae released by adult populations residing in the sandy beach substrate.

Among the decapods, only Luciferidae was recorded sporadically in the winter and summer of the second sampling year, while the larval stages of the other taxa were more abundant during the summer, but constantly present in the plankton throughout the study, with zoeas and megalopae of the infraorder Brachyura being the most frequent in this zone. The occurrence of Decapoda larvae in surf zones has also been described off the coast of the United States, North Carolina (DeLancey, 1987; Stull *et al.*, 2016) and Oregon (Marin Jarrin & Shanks, 2010), and South Africa (Buthelezi, 2012). On the Brazilian coast, Bersano (1994) and Avila *et al.* (2009) also verified the occurrence of Decapoda larvae during the day in the southern region's surf zone, but they did not mention the period of the year in which the catches occurred. According to Resgalla (2001), in coastal areas the occurrence of decapod larvae is related to reproductive peaks where this event is associated with warmer months and greater primary production, thus helping to justify the higher abundances of the group recorded during the summer in this study.

Among the benthoplanktonic organisms, isopods and amphipods were constantly caught throughout the study, regardless of the sampling period, highlighting their importance in the surf zone. These animals are considered important secondary producers and are a source of food for a wide variety of marine predators (Beare & Moore, 1997; MacNeil *et al.,* 1999; Takahashi *et al.*, 1999; Yu *et al.,* 2003).

The Mysidacea were considered to be resident in the upwelling zone, particularly represented by *Metamysidopsis elongata* (with the highest abundance values) and *Bowmaniella brasiliensis* (recorded occasionally and with the lowest abundance). Both species were largely represented by juvenile individuals, followed by adults, comprising males and females that are often ovigerous, reflecting the high recruitment capacity of the Mysida group in turbulent regions (Almeida Prado, 1974; Calil & Borzone, 2008).

Although studies have shown variations in the abundance of misidaceans in relation to the seasons (Almeida Prado, 1974; Mauchline, 1980; Wooldridge, 1981) in this study it was not possible to observe fluctuations in this sense, indicating that in the region the environmental conditions inherent to seasonality may not directly affect the distribution of the group, and that the fauna may be more influenced by the availability of food, predation, salinity and depth (Webb & Wooldridge, 1987; McKenney, 1994).

In the turbulent zone off the Brazilian coast, the occurrence of misidaceans was described by Bersano (1994), where the group was considered to be the main responsible for the abundance and biomass values among the crustaceans. Avila *et al.* (2009) described the adult form of the genus *Metamysidopsis* as the most frequent and abundant in the turbulent zone. Borzone *et al.* (2007), Calil & Borzone (2008), Felipe (2010) and Straub *et al.* (2014) recorded the occurrence of misidaceans on beaches along the coast of Paranà, with the genus *Metamysidopsis* being the most abundant. However, *Bowmaniella brasiliensis* was also recorded by the same authors in the inner subtidal zone, with a lower abundance when compared to the genus *Metamysidopsis*, a similar result in the present study. According to Borzone *et al.* (2007), *Bowmaniella brasiliensis* is more numerous in outer subtidal zones and practically absent in shallow areas, a differential distribution which explains the low density of this species recorded in this study.

The presence of Mysidacea in upwelling zones has also been found in other regions of the world, such as the coast of South Africa (Wooldridge, 1989; Mclachlan & Romer, 1990; Buthelezi, 2012), the west coast of the United States (Clutter, 1967; DeLancey, 1987; Marin Jarrin & Shanks, 2010), Ecuador (Dominguez Granda *et al.*, 2004; Marin Jarrin *et al.*, 2016), Belgium (Lock *et al.*, 1999, 2011; Beyst *et al.*, 2001a, 2001b), Mozambique (Paula *et al.*, 1998), Japan (Takahashi & Kawaguchi, 1997, 1998), France

(San Vicente & Sorbe, 1993) and Argentina (Carcedo *et al.,* 2013). It should be noted, therefore, that among the bentoplankton, the Mysidacea are highly representative in turbulent zones, and can account for approximately 90% of the local biomass (Cockcroft *et al.,* 1988; Mclachlan & Brown, 2006; Bersano, 1994; Buthelezi, 2012).

Misidaceans of the genus *Metamysidopsis* have also been recorded in deeper regions far from the coast by Lopes *et al.* (1986), Fries & Bjornberg (1997) and Resgalla (2001) on the Brazilian coast and by Clutter (1967) in California, showing that although the group is considered to be resident in the upwelling zone, they are not exclusive to this turbulent region. In addition, they can migrate vertically or even distribute and move horizontally during the night (Wooldridge, 1989; Webb & Wooldridge, 1990; Takahashi & Kawaguchi, 1997; Marin Jarrin & Shanks, 2010).

7.2.3. Nictemeral variation

The known nictemeral variation of the zooplankton community described in the literature has not been statistically proven in terms of total abundance, and it is likely that the increase in fauna through the migration of certain organisms during the nocturnal period did not reflect a significant increase in the total abundance of the community in the study area.

However, studies that have analyzed the daily variation of specific groups such as Brachyura, Mysida, Isopoda, Actinopteri in upwelling zones, have described the increase in zooplankton composition and abundance during the nocturnal period, such as research carried out in the North Atlantic (DeLancey, 1987; Lock *et al.,* 1999; Stull *et al.*, 2016), the South Atlantic (Felipe, 2010) and the Pacific Ocean (Takahashi & Kawaguchi, 1997,1998; Marin Jarrin & Shanks, 2010). Thus, if the analysis in this study had been carried out specifically for specific zooplankton groups, the result would possibly have been different, due to the numerically higher catch at night of the holoplankton (Actinopteri), meroplankton (Brachyura) and bentoplankton (Mysida) fauna recorded in this study.

In general, the greater activity of crustaceans in the plankton at night, whether holo, mero or bentoplanktonic organisms, is mainly related to predatory pressure from the daytime fauna, search for food and reproduction (Colman & Segrove, 1955; Romer, 1986; Takahashi & Kawaguchi, 1998; Herd *et al,* 2006; Marin Jarrin & Shanks, 2008; Buthelezi, 2012; Frazào, 2015), which may justify the greater numerical abundance of the Copepoda, Cladocera, Decapoda and Mysida groups at night. Regardless of the reason why these organisms migrate daily, the variation in light intensity is the main stimulus (Gal *et al.*, 1999), however, Melo *et al.* (2010) indicate that other factors can regulate daily migrations, such as the tide, which can alter the composition of the fauna by introducing and leaving coastal organisms and those from distant areas.

The variation in the numerical abundance, frequency of occurrence and percentage contribution of the Actinopteri group, related to the sampling period, was clear for the juvenile stage, with a greater capture of the fauna being recorded at night. Similar results have been described on the coasts of tropical and subtropical regions (Lasiak, 1984; Ruple, 1984; Gibson *et al.,* 1993; Harris, 1996; Godefroid *et al.*, 1998; Layman, 2000; Pessanha & Araùjo, 2003; Pessanha *et al.,* 2003; Marin Jarrin & Shanks, 2010). This pattern can be explained by the nictemeral migration carried out by some species (Methven *et al.,* 2001), which search for food in the water column at night, fleeing from visual predators that are less successful at catching them at night (Mees & Jones 1997; McLachlan & Brown, 2006). However, according to Harris (1996) and Pessanha *et al.* (2003) the ichthyoplankton fauna is also captured less frequently and abundantly during the day due to its visual capacity, which is more efficient during the light period, allowing for greater perception and, consequently, a greater probability of escape at the time of sampling.

The numerical variation of the meroplankton fauna throughout the day was mainly related to the Decapoda groups. All representatives of the Decapoda order were caught during both sampling periods, although with greater numerical abundance, frequency and contribution during the night, with the exception of the carids. Similar results, where greater quantities of decapods (larvae) were recorded during the night, were also observed by DeLancey (1987), Colman & Segrove (1955) and Buthelezi (2012), off the coasts of the United States, England and Africa, respectively. According to

DeLancey (1987), the daily variation of decapods (larvae) in the water column reflects behavioral changes that respond to different stimuli such as light intensity.

Nictemeral migration among the benthoplanktonic fauna was evidenced by the increase in misidaceans in the water column, with these organisms being caught more frequently and abundantly at night. Corroborating this result, Takahashi & Kawaguchi (1997, 1998), Marin Jarrin & Shanks (2010), Melo *et al.* (2010) and Felipe (2010) noted equivalent patterns, where low and zero abundances were obtained during daytime periods, with high abundances in nighttime collections. The greater activity of these crustaceans in the water column at night can be explained by the escape of visual predators (Romer, 1986), since misidaceans are considered the main source of food for fish that frequent the surf zone (Lasiak & Mclachlan, 1987; Marin Jarrin & Shanks, 2008). Migration can also occur for the purposes of mating, feeding and releasing offspring (Clutter, 1967; Stuck *et al.,* 1979; McLachlan, 1983; Takahashi & Kawaguchi, 1997; Heard *et al.,* 2006). In both cases, the variation in light intensity is the main factor responsible for stimulating the migration of many misidaceans (Gal *et al*., 1999).

Despite the greater capture of misidaceans during the night, these organisms were also recorded in this study in daytime collections, where the fauna captured during the day was represented mainly by juveniles, a pattern similar to that presented by Felipe (2010) at Praia do Pontal do Sul/PR. The daily variation of these macroinvertebrates in the water column may be related to their ability to escape during daytime sampling (DeLancey, 1987), thus justifying the capture of young misidaceans, which are less efficient swimmers than adults. However, according to Bersano (1994), the juveniles are distributed more widely within the horizontal patches of misidaceans, occurring more frequently on the margins of the clusters formed by the group, and this formation may explain the occurrence of juveniles closer to the surface due to the shallow depth of the surf zone. In addition, during the daytime, the greater occurrence of juveniles in the water column compared to adults may be related to their smaller size, lower swimming capacity and stabilization close to the substrate, allowing local hydrodynamics to move them more easily to the surface.

With regard to isopods and amphipods, although studies have shown that certain species migrate at night, where during the day they remain close to the substrate, looking for shelter, and at night they move into the water column (Mees & Jones, 1997; McLachlan & Brown, 2006; Marin Jarrin & Shanks, 2010; Melo *et al,* 2010). In the study carried out by Colman & Segrove (1955), they stated that the amphipods in the upwelling zone do not preferentially inhabit the surface layer or the deepest layer, and that the organisms are evenly distributed in the water column, corroborating the data obtained in the present study. Thus, the constant occurrence of the groups in the water column during both periods (day and night) can be explained by the high water dynamics, which maximizes the low swimming and stabilization capacity of these organisms, making it difficult for them to remain close to the substrate, unlike the misidaceans, which have balancing organs that prevent them from being carried away by the action of the currents (Warman *et al.,* 1991; Heard *et al.*, 2006; Borzone *et al.,* 2007; Marin Jarrin & Shanks, 2010). However, Colman & Segrove (1955) also observed, in laboratory experiments, that amphipods make rapid migrations from the bottom to the surface, several times a day and for no apparent reason, which is also a possible reason for the frequent presence of this group in the plankton. This combination of factors could result in the constant occurrence of these individuals, during both periods, in the water column of the surf zone studied. Data corroborating this information was presented by Bersano (1994) and Avila *et al.* (2009), who recorded an abundance of this group during the day on beaches along the southern coast of Brazil.

8. Final considerations

The inner surf zone of Praia da Aviaçào, in Praia Grande/SP, contains a varied set of planktonic organisms. In terms of zooplankton composition, the holoplankton fauna is dominated by small crustaceans from the Copepoda group, with frequent occurrences of Cladocera, as well as Tunicata and Chaetognatha. The meroplankton community is marked by the presence of planktonic larvae of benthic invertebrates from the Mollusca, Polychaeta, Decapoda and Echinodermata (echinopluteus) groups. Among the bentoplankton, the Amphipoda and Mysida groups often appear in the water column. Despite the frequent occurrence of these organisms, the beach analyzed does not show a specific pattern in its zooplankton composition, nor does it represent a closed system, as it is directly influenced by nearby habitats such as rocky shores, due to the constant presence in the samples of larvae from the Hydrozoa, Bryozoa and Cirripedia groups typical of consolidated substrates.

In the area analyzed, there was a high abundance of planktonic organisms with great temporal variability and no direct relationship with seasonal patterns. This high fluctuation in the quantity and composition of zooplankton highlights the dynamic nature of the region, where environmental and biotic processes together can cause recurrent changes in the community structure.

The region also has zooplankton richness and diversity, with these ecological variables showing a slight numerical change in relation to seasonal variability, with the fauna being more heterogeneous in summer compared to winter.

In the turbulent zone analyzed, the nictemeral variation of specific taxonomic groups such as misidaceans, decapods and juvenile fish was observed. However, despite being constant and numerically noticeable, this variation over the course of the day did not statistically reflect a change in the overall abundance of the zooplankton community in the turbulent zone. It is therefore advisable to analyze each group separately in order to really verify the existence of nictemeral variation in the organisms that make up the zooplankton community.

Finally, it can be concluded that the Praia da Aviaçao surf zone is an important habitat for various zooplankton groups, which is evident from the high proportion of planktonic organisms in this turbulent zone, thus indicating the importance of this environment for maintaining the ecological quality of coastal areas, since this group of organisms promotes trophic links that are extremely important for maintaining marine life.

9. Bibliographical references

AAGAARD, T. AND MASSELINK, G. The surf zone. In: SHORT, A. (Ed.). **Handbook of beach and shoreface morphodynamics.** John Wiley & Sons, 1999. p. 72-118.

AB'SABER, A. N. **Litoral do Brasil**. Sao Paulo: Metalivros, 2001.

ALMEIDA PRADO, M. S. Systematics of the Mysidacea (Crustacea) in the Cananéia region. **Boletim do Instituto Oceanogràfico da USP**, v. 23, p. 47-87, 1974.

ALMEIDA, F.F.M. Fundamentos geológicos do relevo paulista. **Geology Bulletin. Sâo Paulo**: Instituto Geogràfico e Geològico, n. 41, p. 169-263, 1964.

ALMEIDA, F.F.M. The system of continental rifts bordering the Santos Basin, Brazil. **Annals of the Brazilian Academy of Sciences**, v. 48, p. 15-26, 1976. Special issue.

ALVARES, C.A.; STAPE, J.L.; SENTELHAS, P.C.; GONÇALVES, J.L.M.; SPAROVEK, G. Koppen's climate classification map for Brazil. **Meteorologische Zeitschrift,** v. 22, n. 6, p. 711-728, 2013.

ALVES, E.C.; PONZI, V.R.A. Morphological sedimentary characteristics of the continental shelf and upper slope of the southeastern continental margin of Brazil. In: BRAZILIAN CONGRESS OF GEOLOGY, 33rd, 1984, Rio de Janeiro. **Annals of Abstracts**. Rio de Janeiro: Sociedade Brasileira de Geologia, v.4., 1984. p.1629-1642.

AMARAL, A.C.Z.; NONATO, E.F. **Anelideos poliquetas da costa brasileira; caracteristicas e chave para familias - Glossàrio**. Brasilia, CNPq/Coordenaçao Editorial, v. 1/2, p.47, 1981.

AMORIM, R.R. **Geoenvironmental analysis with emphasis on the hillside sectors of the urban area of the Municipality of Sâo Vicente-SP.** 2007. 218f. Dissertation (Master's Degree in Geography) - Institute of Geosciences, State University of Campinas, Campinas, 2007.

AMORIM, R.R.; OLIVEIRA, R.C. Anàlise geoambiental com ênfase aos sectores de encosta da àrea urbana do Municipio de Sao Vicente, SP. **Sociedade & Natureza,**

Uberlândia, v.19, p.123-138, 2007.

AMORIM, R.R.; OLIVEIRA, R.C. The study of relief units in municipalities of the Brazilian coastal strip: the example of the Municipality of Sao Vicente - SP. **Caminhos da Geografia**, v.10, n. 30, p. 163-183, 2009.

ANGULO, R.J. Problems in the terminology of classic wave-dominated coastal environments and sub-environments. **Bol. Paranaense Geoc**., v.44, 1996.

ARAÙJO, H.M.P.; RIBEIRO, V.A. Distribution of Chaetognatha species on the continental shelf of Sergipe and Alagoas. **Braz. J. Aquat. Sci. Technol**., v. 9, n.1, p.19-23, 2005.

ATIENZA, D.; CALBET, A.; SAIZ, E.; LOPES, R.M. Ecological success of the cladoceran Penilia avirostris in the marine environment: feeding performance, gross growth efficiencies and life history. **Marine Biology**, v.151, p.1385-1396, 2007.

AVILA, T.R; CATARINA S.; PEDROZO, C.S; BERSANO, J.G. Variaçao temporal do zooplàncton da Praia de Tramandai, Rio Grande do Sul, com ênfase em Copepoda. **Iheringia, Sér. Zool.**, Porto Alegre, v.99, n.1, p.18-26, 2009.

BARREIROS, J.P.; FIGNA, V.; HOSTIM-SILVA, M.; SANTOS, R.S. Seasonal Changes in a Sandy Beach Fish Assemblage at Canto Grande, Santa Catarina, South Brazil. **Journal of Coastal Research,** v.20, n.3, p.862-870, 2004.

BEARE, D.J.; MOORE, P.G. The contribution of Amphipoda to the diet of certain inshore fish species in Kames Bay, Millport. **Journal of Marine Biological Association of the United Kingdom,** v.77, p.907-910, 1997.

BENNETT, B.A. The fish community of a moderately exposed beach on the southwestern cape coast of South Africa and an assessment of this habitat a nursey for juvenile fish. **Est. Coast. Shelf Sci.,** v.28, p:293-305, 1989.

BERSANO, J.C. **Zooplankton of the surf zone of sandy beaches, south of Rio Grande, RS. Spring 1990, summer 1991**. 1994. 163f Dissertation (Master's Degree in Biological Oceanography). Federal University of Rio Grande, Rio Grande. 1994.

BERTOZZI, C.P. **Analysis of artisanal fishing in the region of Praia Grande (SP), period 1999-2001**. 2002. 226f. Dissertation (Master's Degree in Oceanography Biology) - Oceanographic Institute, University of Sao Paulo, Sao Paulo. 2002.

BEYST, B.; CATTRIJSSE, A.; MEES, J. Feeding ecology of juvenile flatfishes of the surf zone of a sandy beach. **Journal of Fish Biology**, v.55, p.1171-1186, 1999.

BEYST, B.; BUYSSE, C.; DEWICKE, A.; MEES, J. Surf zone hyperbenthos of Belgian sandy beaches: Seasonal patterns. **Estuarine, Coastal and Shelf Science,** v.53, n.6, p. 877-895, 2001a.

BEYST, B.; HOSTENS, K.; MEES, J. Factors influencing fish and macrocrustacean communities in the surf zone of sandy beaches in Belgium: temporal variation. **Journal of Sea Research,** v.46, p.281-294, 2001b.

BEYST, B.; HOSTENS, K.; MEES, J. Factors influencing the spatial variation in fish and macrocrustacean communities in the surf zone of sandy beaches in Belgium. **Journal of Marine Biological Association of the United Kingdom**, v.82, p.181187, 2002a.

BEYST, B.; VANAVERBEKE, J.; VINCX, M.; MEES, J. Tidal and diurnal periodicity in macrocrustaceans and demersal fish of an exposed sandy beach, with special emphasis on juvenile plaice *Pleuronectes platessa.* **Marine Ecology Progress Series**, v.225, p.263-274, 2002b.

BIRAJA KU. SAHU; BALIARSINGH, S.K.; SRICHANDAN S.; SAHU, K.C. Zooplankton Abundance and Composition in Surf Zone of Gopalpur Port, East Coast of India-A Case Study. **Marine Science**, v.2, n.6, p.120-124, 2012.

BJORNBERG, T.S.K. Copepoda. In: BOLTOVSKOY, D. (Ed.) **Atlas del Zooplancton del Atlàntico Sudoccidental y métodos de trabajo con el zooplancton marino.** Mar del Plata: Publ. Esp. INIDEP, 1981.v.2, p.587-679 .

BOLTOVSKOY, D. 1981. Data collection and analysis. In: BOLTOVSKOY, D. (Ed.). **Atlas del zooplancton del atlantico sudoccidental y métodos de trabajo con el**

zooplancton marino. Mar del Plata: Publ. Esp. INIDEP, 1981. v.1, p.153168.

BOLTOVSKOY, D. **South Atlantic Zooplankton**. (Ed.) Leinden, The Netherlands: Backhuys Publishers, 1999. v.1/2, p.1706.

BOMTEMPO, V.L. **Hydraulic and sedimentological characteristics of a stretch of the south coast of the State of São Paulo.** 1991. 302f. Dissertation (Master's Degree) - COPPE, Federal University of Rio de Janeiro, Rio de Janeiro. 1991.

BONECKER, A.C.T.; BONECHER, S.L.C.; BASSANI, C. Marine Plankton. In: PEREIRA, R.C.; SOARES-GOMES, A. **Biologia Marinha**. Rio de Janeiro: Interciência. 2002. p.103-125.

BORZONE, C.A.; LORENZI, L.; CALIL, P. Cross-shore Mysidacea (Crustacea) distribution in two sandy beaches with contrasting morphodynamics. **Revista Brasileira de Zoologia**, v.24, n.4, p.943-949, 2007.

BRADFORD-GRIEVE, J.M.; MARKHASEVA, E.L.; ROCHA, C.E.F.; ABIAHY, B. Copepoda. In: BOLTOVSKOY, D. (Ed.) **South Atlantic Zooplankton.** Leiden, Backhuys, 1999. v.2. p.869-1098.

BRANDINI, F.P.; LOPES, R.M.; GUTSEIT, K.S.; SPACH, H.L.; SASSI, R. **Planktonology on the Brazilian continental shelf. Diagnosis and Bibliographic Review.** Rio de Janeiro: REVIZEE. MMA, CIRM, FEMAR, 1997. p. 196.

BROWN, A.C.; McLACHLAN, A. **Ecology of sandy shores**. New York: Elsevier, 1990.

BUTHELEZI, P.P. **Invertebrate Plankton Assemblages In The Surf Zones Of The St. Lucia And Mhlathuze Estuaries, Kwazulunatal, South Africa**. 2002. 152f. Dissertation (Master of Science) - Department of Zoology, University of Zululand, South Africa. 2002.

CALIL, P.; BORZONE, C.A. Population structure and reproductive biology of *Metamysidopsis neritica* (Crustacea: Mysidacea) in a sand beach in South Brazil. **Revista Brasileira de Zoologia**, v. 25, n.3, p.403-412, 2008.

CALLIARI, L.J.; MUEHE, D.; HOEFEL, F.G.; TOLDO, E. Beach morphodynamics: a brief review. **Rev. Bras. Oceanogr.**, v.51, p.63-78, 2003.

CAMPBELL, E.E.; BATE, G.C. Ground water in the Alexandria dune field and its potential influence on the adjacent surf-zone. **Water AS**, v.17, n.2, p.155-160, 1991a.

CAMPBELL, E.E.; BATE, G.C. The Flora of the Sandy Beaches of Southern Africa. III. The South Coast Microflora. **Institute for Coastal Research Report,** n.26, p.129, 1991b.

CAMPBELL, E.E.; BATE, G.C. Physical features of the South African coastline related to the occurrence of surf diatom accumulations. **Botanica Marina**, v.40, p.179-185, 1997.

CAMPOS, M.A.G. The **Appendicularia (Chordata: Tunicata) of the region between Cabo Frio (RJ) and Cabo de Santa Marta Grande (SC).** 2000. 120f. Dissertation (Master's Degree in Biological Oceanography) - Istituto

Oceanogràfico, University of Sao Paulo, Sao Paulo. 2000.

CAMPOS, M.A.G. **Appendiculariae from the Brazilian coast**. 2004. 169f. Thesis (Doctorate in Science, Biological Oceanography) - Oceanographic Institute, University of Sao Paulo, Sao Paulo. 2004.

CARCEDO, C.; FIORI, S.; BREMEC, C. Macrobenthic surf zone communities of temperate sandy beaches: spatial and temporal patterns. **Marine Ecology Progress Series**, p.1-11, 2014.

CARTER, R.W.G. **Coastal environments: an introduction to the physical, ecological and cultural systems of coastlines**. London: Academic Press, 1988.

CAVALCANTI, E.A.H.; LARRAZABAL, M.E.L. Macrozooplàncton of the Exclusive Economic Zone of Northeastern Brazil (Second Oceanographic Expedition - REVIZEE/NE II) with emphasis on Copepoda (Crustacea). **Revista Brasileira de Zoologia,** v.21, n.3, p.467-475, 2004.

CLARK, B.M.; BENNETT, B.A.; LAMBERTH, S.J. Temporal variations in surf zone fish assemblages from False Bay, South Africa. **Marine Ecology Progress Series**, v.131, p.35-47, 1996a.

CLARK, B.M.; BENNETT, B.A.; LAMBERTH, S.J. Factors affecting spatial variability in seine net catches of fish in the surf zone of False Bay, South Africa. **Marine Ecology Progress Series**, v.131, p.17-34, 1996b.

CLARK, B.M. Variation in surf-zone fish community structure across a wave exposure gradient. **Estuarine, Coastal and Shelf Science**, v. 44, p. 659-674, 1997.

CLUTTER, R.J. Zonation of nearshore mysids. **Ecology**, v.48, p.200-208, 1967.

COCKCROFT, A.C.; McLACHLAN, A. Distribution of juvenile and adult penaeid prawns Macropetasma africanus (Balss) in Algoa Bay. S. Afr. **Journal of Marine Science**, v.4, p.245-255, 1986.

COCKCROFT, A.C.; WEBB, P.; WOOLDRIDGE, T. Nitrogen regeneration by two surf-zone mysids, *Mesopodopsis slabberi* and *Gastrosaccus psammodytes* **Marine Biology**, v.99, p.75-82, 1988.

COLMAN, J.S.; SEGROVE, F. The tidal plankton over Stoupe Beck sands, Robin Hood's bay (Yorkshire, North Riding). **Journal of Animal Ecology**, v.24, p.445462, 1955.

CONTI, J.B.; FURLAN, S.A. Geoecology: climate, soils and biota. In: ROSS, J.L.S. (Ed.) **Geography of Brazil**. Sao Paulo: Edusp, 2011. p.67-208.

COSTA, K.G.; PINHEIRO, P.R.S.; MELO, C.A.R.; OLIVEIRA, S.M.O.; PEREIRA, L.C.C.; COSTA, R.M. . Effects of seasonality on zooplankton community dynamics in the macrotidal coastal zone of the Amazon region. **Journal of Coastal Research,** v. 64, p.364-368, 2011.

COSTA, D.; LEITE, N.R.; ABREU, R.; FERNADES, A.; FERNANDES, G.; BARROS, R. Distribution, occurrence and composition of zooplankton at Ajuruteua Beach, Bragança-PA. In: BRAZILIAN CONGRESS OF OCEANOGRAPHY, 2014, Itajai-SC. **Annals of Abstracts**. Brazilian Association of Oceanography, 2014a. p.1110-1111.

COSTA, D.; LEITE, N.R.; ABREU, R.; FERNADES, A.; FERNANDES, G.; BARROS, R. Distribution, occurrence and composition of copepods at Ajuruteua Beach, Bragança-PA. In: BRAZILIAN CONGRESS OF OCEANOGRAPHY, 2014, Itajai-SC. **Annals of Abstracts**. Brazilian Association of Oceanography, 2014b. p.1112-1113.

CRAWLEY, K.R.; HYNDES, G.A.; AYVAZIAN, S.G. Influence of different volumes and types of detached macrophytes on fish community structure in surf zones of sandy beaches. **Marine Ecology Progress Series,** v.307, p.233-246, 2006.

DAVIDSON-ARNOTT, R. **An Introduction to Coastal Processes and Geomorphology**. England: Cambridge University Press, 2010.

DAVIS, R.A.; HAYES, M.O. What is a wave-dominated coast? **Marine Geology**, v.60, p.313-329, 1984.

DEFEO, O.; MCLACHLAN, A.; SCHOEMAN, D.S.; SCHLACHER, T.A.; DUGAN, J.; JONES, A.; LASTRA, M.; SCAPINI, F. Threats to sandy beach ecosystems: A review. **Estuarine, Coastal and Shelf Science,** v.81, p.1-12, 2009.

DELANCEY, L. The summer zooplankton of the surf zone at folly beach, South Carolina. **Journal of Coastal Research**, v.3, p.211-217, 1987.

DEXTER, D.M. Temporal and spatial variability in the community structure of the fauna of four sandy beaches in South-eastern New South Wales. **Australian Journal of Marine & Freshwater Research**, v. 34, p. 663-672, 1984.

DOMINGUEZ-GRANDA, L.; FOCKEDEY, N.; DE-MEY, M.; BEYST, B.; CORNEJO, M.; CALDERON, J.; VINCX, M. Spatial patterns of the surf zone hyperbenthic fauna of Valdivia Bay (Ecuador). **Hydrobiologia**, v.529, p.205-224, 2004.

DYE, A.H.; MCLACHLAN, A.; WOOLDRIDGE, T. The ecology of sandy beaches in Natal. ***South African Journal*** of ***Zoology***, v. 16, n.4, p. 200-209, 1981.

ECHER, M.P.; MARTINS, F.R; PEREIRA, E.B. The importance of cloud cover data and its variability: methodologies for data acquisition. **Revista Brasileira de Ensino**

de Fisica, v. 28, n. 3, p. 341-352, 2006.

ESNAL, G.B. Appendicularia. In: BOLTOVSKOY, D. **South Atlantic Zooplankton** (Ed.). Leinden, The Netherlands: Backhuys Publishers, 1999. v.1, p.1375-1399.

FARINACCIO, A. **Impacts on coastal dynamics resulting from interventions on sandy beaches and estuarine channels of densely occupied areas on the coast of Sao Paulo, an application of knowledge to unoccupied areas**. 2008. 229f. Thesis (PhD in Chemical Oceanography and Geology) - Oceanographic Institute, University of Sao Paulo, Sao Paulo. 2008.

FAUCHALD, K.; JUMARS, P. The diet of worms: a study of polychaeta feeding guilds. **Oceanography and Marine Biology, An Annual Review,** v.17, p.193284, 1979.

FAVERO, J.M.; DIAS. J.F. Spatio-temporal variation in surf zone fish communities at Ilha do Cardoso State Park, Sao Paulo, Brazil. **Latin American Journal of Aquatic Research**, v. 41, n. 2, p. 239-253, 2013.

FAVERO, J. M.; DIAS. J. F. Juvenile fish use of the shallow zone of beaches of the Cananéia-Iguape coastal system, southeastern Brazil. **Brazilian Journal of Oceanography**, v.63, n.2, p.103-114, 2015.

FELIPE, F.P. **Nictemeral variation of misids (Crustacea, Peracarida) in the surf zone of Pontal do Sul Beach - PR**. 2010. 52f. Course Conclusion Paper (Graduation in Oceanography) - Institute of Science and Technology, Federal University of Paranà, Paranà. 2010.

FERREIRA, L.C. **O fitoplâncton na zona de arrebentaçâo de praias urbanas da regiâo metropolitana do recife, Pernambuco, Brasil.** 2007. 147f. Dissertation (Master's Degree in Oceanography) - Federal University of Pernambuco, Pernambuco 2007.

FOLK, R.L.; WARD, W.C. Brazos river bar: A study in the significance of grain size parameters. **Journal of Sed. Petrol**, v.27, n.1, p.3-26, 1957.

FONTES, R. **Currents in the São Sebastiâo Channel.** 1995. 159f. Dissertation

(Master's Degree in Physical Oceanography) - Oceanographic Institute, University of Sao Paulo, Sao Paulo. 1995.

FONTES, R. **Water masses and ocean currents: Coastal and marine waters.** In: Sao Paulo (State). Secretariat for the Environment. 2010. Xixovà-Japui State Park - Management Plan - Main Volume. Sao Paulo: Coordination of Technical Information, Documentation and Environmental Research; Forestry Institute; Botanical Institute; CEPEL-UNESP, 544p. (Environmental Documents Series).

FONZAR, B.C. Atmospheric circulation in South America: the major planetary systems and regional subsystems that affect the continent (location and trajectories). **Caderno de Geociências**, Rio de Janeiro, v.11, p.11-33, 1994.

FRAZÂO, M.A.D. **Surf zone zooplankton communities spatial patterns on rocky and sandy beaches of the Portuguese coast**. 2015. 50f. Dissertation (Master in Ecology) - Department of Life Sciences, Faculty of Science and Technology, University of Coimbra, Portugal. 2015.

FRIES, B.G.; BJORNBERG, T.K.S. Aspects on the biology and ecology of the misidaceans of Sao Sebastiao, SP, Brazil. In: SIMPÒSIO DE BIOLOGIA MARINHA, 11, Sao Sebastiao, 1996.

FÙLFARO, V.J.; COIMBRA, A.M. The beaches of the São Paulo coast. In: BRAZILIAN CONGRESS OF GEOLOGY, XXVI, SBG, Belém. Abstracts of Communications. Belém, 1972. p. 253-255.

FÙLFARO, V.J.; PONÇANO, W.L. Recent tectonic features in the Serra do Mar region, and its importance to engineering geology. In: I.A. E. G. CONGRESS, II, 1974, Sao Paulo. Proceedings, Sao Paulo. v. I, p. II-7.1 to II-7.5.

FÙLFARO, V.J.; SUGUIO, K. The genesis of the São Paulo coastal plains. In: BRAZILIAN CONGRESS OF GEOLOGY, 28. Porto Alegre-RS, Anais, Porto Alegre - RS, 1974. p.37-42.

FÙLFARO, V.J.; SUGUIO, K.; PONÇANO, W.L. A Gênese das Planicies Costeiras Paulistas. In: BRAZILIAN CONGRESS OF GEOLOGY, 28, 1974. Porto Alegre -

Proceedings. Porto Alegre: SBG, v. 3, p. 37-42.

FURTADO, W; MAHIQUES, M.M. Sediment distribution in coastal regions and the northern continental shelf of the State of São Paulo. In: SIMPÒSIO DE ECOSSISTEMAS DA COSTA SUL E SUDESTE BRASILEIRA, ESTRUTURA, FUNÇÂO E MANEJO, 2., 1990, Aguas de Lindóia. Publi. AC. Ciênc. Est. S. Paulo v.71, n.1, p.20-29.

GAELZER, L.R.; ZALMON, I.R. The influence of wave gradient on the ichthyofauna of southeastern brazil: focusing the community structure in surf zone. **Journal of Coastal Research**, v.35, p.456-462, 2003.

GIANNINI, R.; PAIVA-FILHO, A.M. Comparative analysis of the ichthyofauna of the surf zone of sandy beaches on the coast of the State of Sao Paulo, Brazil. **Bol. Inst. Ocean.,** v.43, p.141-152, 1995.

GIANUCA, N. M. A preliminary account of the ecology of sandy beaches in southern Brazil. In: McLACHLAN. A.; ERASMUS, T. **Sandy Beaches as Ecosystems**. (1th ed.). The Hague: Dr. Junk Publishers, 1983. p.413-420.

GIBSON, R.N.; ANSELL, AD.; ROBB, L. Seasonal and annual variations in abundance and species composition of fish and macrocrustacean communities on a Scottish sandy beach. **Marine Ecology Progress Series,** v.98, p.89-105, 1993.

GODEFROID, R.S.; HOFSTAETTER, M.; SPACH, H.L. Moon, tidal and diel influences on catch composition of fishes in the surf zone of Pontal do Sul beach, Paranà. **Ver Bras Zool,** v.15, p.697-701, 1998.

GODEFROID, R.S.; SPACH, H.L.; MACLAREN, G.; SCHWAIZ JR., R. Temporal changes in the abundance of the fish fauna of the shallow infralittoral of a beach, southern Brazil. **Iheringia, Sér. Zool**., v.94, p.95-104, 2004.

GUEDES, C.C.F. **Evoluçâo Sedimentario Quaternària da Ilha Comprida, Estado de Sâo Paulo.** 2009. 147f Dissertation (Master's Degree) - Institute of Geosciences, University of Sao Paulo, Sao Paulo. 2009.

GUERRA, A.J.; GUERRA, A.J.T. Dicionàrio geológico-geomorfológico. 5th edition. Rio de Janeiro: Betrand Brasil. 2005.

HAMERLYNCK, O.; MEES, J. Temporal and spatial structure in the hyperbenthic community of a shallow coastal area and its relation to environmental variables. **Oceanologica Acta** (Special Volume), v.11, p. 205-212, 1991.

HANEY, J.F.; HALL, D.J. Sugar-coated Daphnia: A preservation technique for Cladocera. **Limnology and Oceanography**, v.18, p.331-333, 1973.

HARARI, J.; CAMARGO, R. Estimation of the volume transport by the tides in the Southeastern Brazilian Platform through a hydrodynamical numerical model. In: SIMPÓSIO DE ECOSSISTEMAS DA COSTA SUL E SUDESTE BRASILEIRA: ESTRUTURA, FUNÇÂO E MANEJO, 2., 1990, Aguas de Lindóia - SP. v.1, p.75-83.

HARARI, J.; FRANÇA, C.A.S.; CAMARGO, R. Climatology and Hydrography of Santos Estuary. In: NEVES, R; BARETTA, J.; MATEUS, M. (Orgs.). **Perspectives on Integrated Coastal Zone Management in South America.** Portugal: IST Press, 2008. p.147-160.

HARRIS, S.A. **Larval fish assemblages of selected estuarine and coastal systems in Kwazulu-Natal, South Africa**. 1996. 241p. Ph.D. Thesis (Doctorate in Philosophy) - Department of Zoology, University of Zululand, South Africa. 1996.

HARRIS, S.A.; CYRUS, D.P. Larval and juvenile fishes in the surf zone adjacent to the St. Lucia Estuary Mouth, Kwazulu-Natal, South Africa. **Marine and Freshwater Research**, v.47, p. 465-482, 1996.

HEARD, R.W.; PRICE, W.W.; KNOTT, D.M.; KING, R.A.; ALLEN, D.M. **Taxonomic Guide of the Mysids of the South Atlantic Bight**. (Ed.) NOAA Professional Paper NMFS 4. US Department of Commerce, 2006. p. 1-37.

HEWSON, I.; O'NEIL, J.M.; ABAL, E. A low-latitude bloom of the surf-zone diatom, *Anaulus australis* (Centrales, Bacillariophyta) on the coast of Southern Queensland (Australia). **Journal of Plankton Research,** v.23, p. 1233-1236, 2001.

HOEFEL, F.G. **Morphodynamics of sandy ocean beaches: a bibliographical review.** (Ed.) Univali, Itajai, 1998. p.92.

HUYS, R.; BOXSHALL, G.A. **Copepod evolution**. London: The Ray Society, 1991. v.159, p.468.

IPT - Institute for Technological Research. **Survey of the physical environment and establishment of normative criteria for the urban occupation of the hills of Santos and São Vicente.** Sao Paulo (Report n. 11.599). 1981.

JOHNSON W.S.; ALLEN D.M. **Zooplankton of the Atlantic and gulf coasts: a guide to their identification and ecology**. Baltimore: The Johns Hopkins University Press, 2005. p.379.

KAHN, A.E.; CAHOON, L.B. Phytoplankton productivity and protophysiology in the surf zone of Sandy beaches in North Carolina, USA. **Estuarics and Coasts**, v. 35, p.1393-1400, 2012.

KOMAR, P.D. **Handbook of coastal processes and erosion** (4th ed.). CRC Press, p.297, 1991.

KREBS, C.J. **Ecological Methodology**. New York: Harper-Collins Publishers, 1989. p.654.

KURTZ, F.W. **Larval dynamics of *Sardinella brasiliensis* (Steindachner, 1879) (Teleostei, Clupeidae) in Southeast Brazil and implications for recruitment.** 1999. 169f. Thesis (Doctorate in Sciences, Area Oceanography Biology) - Oceanographic Institute, University of Sao Paulo, Sao Paulo. 1999.

LAMPARELI, C. Mapping the coastal ecosystems of the State of São Paulo. Sao Paulo: Secretaria do Meio Ambiente/ CETESB. p.108, 1998.

LASIAK, T.A. Nursey grounds of juvenile teleosts: evidence from the surf-zone of King's Beach, Port Elizabeth. **African Journal of Marine Science**, v.77, p.388390, 1981.

LASIAK, T.A. Structural aspects of the surf-zone fish assemblage at King's Beach, Algoa Bay, South Africa: short-term fluctuations. **Estuarine, Coastal and Shelf Science**, v.18, n.3, p.374-360, 1984.

LASIAK, T.A. Juveniles, food and the surf zone habitat implications for teleost nursery areas. South **African Journal of Zoology**, v.21, n.1, p.51-56, 1986.

LASIAK, T.; McLACHLAN, A. Opportunistic Utilization of Mysid Shoal by Surfzone Teleosts. **Marine Ecology Progress Series**, v.37, p.1-7, 1987.

LAYMAN, C.A. Fish assemblage structure of the shallow ocean surf-zone on the eastern shore of Virginia Barrier Islands. **Estuarine, Coastal and Shelf Science,** v.51, n.2, p.201-213, 2000.

LENZ, J. Introduction. In: HARRIS, R.P.; WIEBE, P.H.; LENZ, J.; SKJOLDAL, H.R., HUNTLEY, M. (Ed.), **Zooplankton Methodology Manual**. Amsterdam: Elsevier, 2005. p.1-32.

LEVINTON, J.S. **Marine Ecology**. Prentice-Hall Inc., Englewood Cliffs. 1982. p.526.

LEWIN, J.; SCHAEFER, C.T. The role of phytoplankton in surf ecosystems. In: McLACHLAN, A.; ERASMUS, T. In: McLACHLAN. A.; ERASMUS, T. **Sandy Beaches as Ecosystems**. (1th ed.). The Hague: Dr. Junk Publishers, 1983. p.381390.

LIANG, T.H.; VEGA-PEREZ, L.A. Studies on Chaetognaths off Ubatuba Region, Brazil. I. Distribution and abundance. **Bolm Inst. oceanogr.**, S. Paulo, v.42, n.1-2, p.73-84, 1994.

LIPPMANN, T.C.; HOLMAN, R.A. The spatial and temporal variability of sand bar morphology. **Journal of Geophysical Research**, v.95, n.7, p.1575-11590, 1990.

LOCK, K.; BEYST, B.; MEES, J. Circadiel patterns in the tidal plankton of a sandy beach in Zeebrugge (Belgium). **Belgian Journal of Zoology,** v.129, p.339-352, 1999.

LOCK, K.; MEES, J.; VINCX, M.; GOETHALS, P.L.M. Did global warming and alien invasions affect surf zone hyperbenthic communities on sandy beaches in Belgium? **Hydrobiologia**, v.664, p.173-181. 2011.

LOPES, R.M.; ALMEIDA PRADO-POR, M.S.; POR, F.D. Zooplankton seasonality in the Rio Verde estuary (Juréia, Sao Paulo, Brazil). **Revista de Hidrobiologia Tropical**, v. 19, p.207-214, 1986.

LOPES, R.M.; MONTÉ, M.A.; GORRI, C.; MUXAGATA, E.; MIYASHITA, L.; OLIVEIRA, L.P. The marine zooplankton of the region between Cabo de Sao Tomé (RJ) and Chui (RS). In: ROSSI-WONGTSCHOWSKI, C.L.D.B.; MADUREIRA, L.S. **O ambiente oceanogràfico da plataforma continental e do talude na regiâo sudeste-sul do Brasil** (Ed.). Sao Paulo: EDUSP, 2006. p.265-358.

LOPES, R.M. Marine zooplankton studies in Brazil: A brief evaluation and perspectives. **Anais da Academia Brasileira de Ciências**, v.79, n.3, p.369-379, 2007.

MAcMAHAN, J.H.; BROWN, J.; THORNTON, E.B.; RENIERS, A.J.H.M.; STANTON, T.P.; HENRIQUEZ, M.; GALLAGHER, E.; MORRISON, J.; AUSTIN, M.; SCOTT, T.; SENECHAL, N. Mean Lagrangian flow behavior on an open coast rip-channeled beach: new perspectives. **Marine Geology**, v.268, p.115, 2009.

MACNEIL, C.; DICK, J.T.A.; ELWOOD, R.W. The dynamics of predation on *Gammarus* spp. (Crustacea: Amphipoda). **Biol Rev,** v.74, p.375-395, 1999.

MARAZZO, A.; NOGUEIRA, C.S.R. Composition, spatial and temporal variations of Chaetognatha in Guanabara Bay, Brazil. **Journal of Plankton Research**, v.18, n.12, p.2367-2376, 1996.

MARIANI, S. Can spatial distribution of ichthyofauna describe marine influence on coastal lagoons? A central Mediterranean case study. **Estuarine, Coastal and Shelf Science**, v.52, p.261-267, 2001.

MARIN JARRIN, J.R.; SHANKS, A.L. Spatio-temporal dynamics of the surf-zone faunal assemblages at a Southern Oregon sandy beach. **Marine Ecology**, v.32, p.232-242, 2010.

MARIN JARRIN, J.R.; MINO QUEZADA, S.L.; DOMINGUEZ-GRANDA, L.E.; ARGUDO, S.M.G.; GRUNAUER, M.P.C.R. Spatio-temporal variability of the surf-zone fauna of two Ecuadorian sandy beaches. **Marine and Freshwater Research**, v.67, p.566-577, 2016.

MARTINELLI FILHO, J. E. **Temporal variation and growth of zooplankton on the north coast of São Paulo, with emphasis on immature stages of copepods**. 2013.

261f. Thesis (Doctorate in Sciences, Biological Oceanography Area) - Oceanographic Institute, University of Sao Paulo, Sao Paulo. 2013.

MATOS, J.B.; SILVA, N.I.S.; PEREIRA, L.C.C.; COSTA, R.M. Quali-quantitative characterization of the phytoplankton of the upwelling zone of an Amazonian beach. **Acta Botanica Brasilica**, v.26, n.4, p.979-990, 2012.

MAUCHLINE, J. The biology of mysids and euphausiids. **Advances in Marine Biology,** p.681, 1980.

MCKENNEY, C.L. Resistance patterns to salinity and temperature in an estuarine *mysid* (*Mysidopsis bahia*) in relation to its life cycle. **Comparative Biochemistry and Physiology, Part A. Physiology, v.109, n.**1, p. 199-208, 1994.

McLACHLAN, A. The definition of sandy beaches in relation to exposure: a simple system. **South African Journal of Science**, v.76, p.137-138, 1980.

McLACHLAN, A.; ERASMUS, T.; VAN-DER HORST, G.; ROSSOUW, G.; LASIAK, T.A.; McGMYNNE, L. Sand beach energetics: an ecosystem approach towards a high energy interface. **Estuarine, Coastal and Shelf Science**, v.13, p.11-25, 1981.

McLACHLAN, A.; HESP, P. Faunal response to morphology and water circulation of a sandy beach with cusps. **Marine Ecology**, v.19, p. 133-144, 1984.

McLACHLAN, A. Sandy beach ecology: A review. In: McLACHLAN, A.; ERASMUS, T. In: McLACHLAN. A.; ERASMUS, T. **Sandy Beaches as Ecosystems**. (1th ed.). The Hague: Dr. Junk Publishers, 1983. p.321-380.

McLACHLAN, A.; ERASMUS, T. (1st ed.). **Sandy beaches as ecosystems**. The Hague: Dr. Junk Publishers, 1983. p.756.

McLACHLAN A. Dissipative Beaches and Macrofauna Communities on Exposed Intertidal Sands. **Journal of Coastal Research**, v.6, n.1, p.57-71, 1990.

McLACHLAN, A.; ROMER, G. Trophic Relationships in a High Energy Beach and Surf-zone Ecosystem. In: BARNES, M.; GIBSON, R. (Ed.). **Trophic relations in the**

marine environment. Aberdeen: Univ. Press.1990. p.365.

McLACHLAN, A.; DORVLO, A. Global patterns in sandy beach macrobenthic communities. **Journal Coastal Research**, v.21, n.4, p.674-687, 2005.

McLACHLAN, A.; BROWN, A. **The Ecology of Sandy Shores**. (2th ed.) Academic Press. 2006.

MEES, J.; JONES, M.B. The Hyperbenthos. **Oceanography and Marine Biology**: an Annual Review, v.35, p.221-255, 1997.

MELO, M.S.; PONÇANO, W.L. Genesis, distribution and stratigraphy of the Cenozoic deposits of the State of São Paulo. **Sâo Paulo: Instituto de Pesquisas Tecnológicas**. v.9, p.1-21 (Monograph Series), 1983.

MELO, P.A.M.C.; SILVA, T.A.; NEUMANN-LEITÂO, S.; SCHWAMBORN, R.; GUSMÂO, L.M.O.; PORTO-NETO, F. Demersal zooplankton communities from tropical habitats in the southwestern Atlantic. **Marine Biology Research,** v.6, p.530-541, 2010.

MENEZES, M.P. **Reconstruction of the coastal sedimentary dynamics of the Juréia-Itatins Ecological Station (SP) and its correlation with local human activity**. 1994. 109f. Dissertation (Master's Degree in Environmental Science) - PROCAM, University of Sao Paulo, Sao Paulo. 1994.

MESQUITA, A.R. Mean sea level variations on decadal and interdecadal scales off the southeast coast of Brazil. In: SEMINAR ON OCEAN WAVES AND TIDES. ARRAIAL DO CABO (RJ), 1995, Arraial do Cabo. **Abstract Proceedings,** 1995. p. 52-53.

METHVEN, D.A.; HAEDRICH, R.L.; ROSE, G.A. The fish assemblage of a Newfoundland estuary: daily, monthly and annual variation. **Estuarine, Coastal and Shelf Science,** v.52, p. 669-687, 2001.

MILLER, C.B. **Biological Oceanography**. Malden: Blackwell Science, 2004. p.402.

MIYASHITA, L.K.; GAETA, S.A.; LOPES, R.M. Life cucle and reproductive traits of marne podonids (Cladocera, Onychopoda) in a coastal subtropical area. **Journal of Plankton Research**, v.33, n.5, p.779-792, 2011.

MIYASHITA, L.K.; POMPEU, M.; GAETA, S.A.; LOPES, R.M. Seasonal contrasts in abundance and reproductive parameters of *Penilia avirostris* (Cladocera, Ctenopoda) in a coastal subtropical area. **Marine Biology**, v.157, p.2511-2519, 2010.

MODDE, T. Growth and residency of juvenile fishes within a surf zone habitat in the gulf of México. **Gulf Res. Rep**., v.6, n.4, p.377-385, 1980.

MONTEIRO, C.A.F. A Dinàmica Climàtica e as Chuvas no Estado de Sao Paulo: estudo geogràfico sob forma de atlas. Sao Paulo: University of Sao Paulo/ Institute of Geography, 1973. p 129.

MONTEIRO-NETO, C.; MUSICK, J.A. Effects of beach seine size on the assessment of surf-zone fish communities. **Atlantica,** v.16, p. 23-29, 1994.

MONTEIRO-NETO, C.;. CUNHA, L.P.R.; MUSICK, J.A. Community structure of surf-zone fishes at Cassino Beach, Rio Grande do Sul, Brazil. **Journal of Coastal Research**, v.35, p.492-501, 2003.

MORAN, S. Ecology of distribution of the sand-dwelling mysid, *Gastrosaccus sanctus* (Van Beneden 1961) along the Mediterranean sandy shore of Israel. **Crustraceana Supplement**, v.3, p.357-361, 1972.

MOURA, M.R. **Coastal dynamics and vulnerability to coastal erosion in the municipalities of Caucaia and Aquiraz, Ceará. 2012**. 210f. Thesis (PhD in Geography) - Department of Geography, Federal University of Ceará, Ceará. 2012.

MUEHE, D. 1994. Coastal Geomorphology. In:. GUERRA, A.J.T.; CUNHA, S.B. (Orgs) Geomorfologia: Uma actualizaçao de Bases e Conceitos. Rio de Janeiro: Bertrand Brasil S.A. 1994.

NEWELL, G.E.; NEWELL, R.C. Marine Plankton: a practical guide. London: Hutchinson, 1963. p. 221.

NIBAKKEN, J.W. Marine Biology: an ecological approach (3rd Ed.). New York:

Harper Colling College Publishers, 1993. p.462.

NIELSEN T. G., MUNK P. Zooplankton diversity and the predatory impact by larval and small juvenile fish at the Fisher Banks in the North Sea. **Journal of Plankton Research**, v.20, p.2313-2332, 1998.

NIMER, E. Climatology of Brazil. Rio de Janeiro: IBGE (2 Ed.), 1979, p. 421.

NITTROUER, C.A.; WRIGHT, L.D. Transport of particles across continental shelves. **Geophysical Research**, v.32, n.1, p.85-113, 1994.

OMORI, M. & IKEDA, T. Methods in marine zooplankton ecology. New York: John Wiley. 1984. p.332

PAIVA-FILHO, A.M. & TOSCANO, A.P. Comparative study and seasonal variation of the ichthyofauna in the intertidal zone of Mar Casado-Guarujà and Mar Pequeno-Sao Vicente, SP. **Bol. Inst. Ocean.**, v.35, p.153-165, 1987.

PARSONS, T.R.; TAKAHASHI, M.; HARGRAVE, B. Biological oceanographic processes (3rd Ed.). Pergamon Press, Oxford. 1984. p.332

PAULA, J.; PINTO, I.; GAUMBE, I.; MONTEIRO, S.; GOVE, D.; GUERREIRO. J. Seasonal cycle of planktonic communities at Inhaca Island, southern Mozambique. **Journal of Plankton Research,** v.20, n.11, p.2165-2178, 1998.

PEREIRA, J.B. 2010. Composition, distribution, biomass and secondary production of zooplankton in the Santos Estuarine System, São Paulo, Brazil. 2010. 303f. Thesis (Doctorate in Science, Biological Oceanography) - Oceanographic Institute, University of Sào Paulo, Sào Paulo. 2010.

PESSANHA, A.L.M.; ARAÙJO, F.G.; AZEVEDO, M.C.A.; GOMES, I.D. Spatial and temporal variations in the composition and structure of the young fish community of Sepetiba Bay, RJ. In: ENCONTRO BRASILEIRO DE ICTIOLOGIA, SÀO CARLOS. Annals of Abstracts. Sào Carlos: Federal University of Sào Carlos, 1999. p.671.

PESSANHA, A.L.M.; ARAÙJO, F.G.; AZEVEDO, M.C.C.; GOMES, I.D. Variaçôes temporal e espaciais da comunidade de peixes jovens da Baia de Sepetiba, Rio de Janeiro. **Revista Brasileira de Zoologia**, v.17, p.251-261, 2000.

PESSANHA, A.L.M.; ARAÙJO, F.G. Spatial, temporal and diel variations of fish assemblages at two Sandy beaches in the Sepetiba Bay, Rio de Janeiro, Brazil. **Estuarine, Coastal and Shelf Science**, v.57, p,817-828, 2003.

PESSANHA, A.L.M., ARAÙJO, F.G., AZEVEDO, M.C.C. & GOMES, I.D. 2003. Diel and seasonal changes in the distribution of fish on a southeast Brazil sandy beach. **Marine Biology**, v.143, p.1047-1055, 2003.

PIELOU, E.C. **Mathematical ecology**. New York: Jon Wiley & Sons, 1977. p. 385.

PINHEIRO, S.C.C.; LEITE, N.R.; COSTA, V.B.; COSTA, K.G.; PEREIRA, L.C.C.; COSTA, R.M. Spatial-temporal influence of hydrological variables on the diversity and abundance of copepods on an equatorial macrotidal beach in the Brazilian Amazon region. **Journal of Coastal Research**, v.64, p. 425-429, 2011.

PONÇANO, W.L. General Characteristics of Sedimentation and Coastal Currents between Santos and Cananéia. **Noticia Geomorfológica**, Campinas, v.16, n.31, p.107111, 1976.

POTTER, I.C.; BIRD, D.J.; CLARIDGE, P.N.; CLARKE, K.R.; HYNDES, G.A.; NEWTON, L.C. Fish fauna of the Severn Estuary. Are there longterm changes in abundance and species composition and are the recruitment patterns of the main species correlated? **Marine Ecology Progress Series**, v.258, p.15-37, 2001.

PRAIA GRANDE CITY HALL. Available at: <http://www.praiagrande.sp.gov.br/Atendimento/index-.asp?cd_pagina=75>.

Accessed on: July 23, 2015.

RAYMONT, T.E. Plankton and productivity in the oceans. Zooplankton. Oxford, Pergamon Press. v.2. 1983.

RESGALLA, J.R.C. 2001. Study of the environmental impact on the Zooplankton community in the saco dos limões cove, southern bay of the island of Santa Catarina,

Brazil. **Atlântica**, v.23, p.5-16, 2001.

RICKLEFS, R.E. **The Economy of Nature**. Rio de Janeiro: Editora Guanabara, 1996. p.470.

RODRIGUES, M.; FURTADO, V.V.; TESSLER, M.G.; E MAHIQUES, M.M. **Atlas sedimentológico da plataforma continental do Estado de Sâo Paulo: texto explicativo e mapas**. Sâo Paulo: IOUSP. 2003. p 34.

ROLIM, G.S., CAMARGO, M.P.B.; LANZA, D.G.; MORAES, J.F.L. Koppen and Thornthwaite climate classifications and their applicability in determining agroclimatic zones for the state of Sâo Paulo. - **Bragantia**, v.4, p.711-720, 2007.

ROMER, G.S. Faunal assemblages and food chains associated with surf zone phytoplankton blooms. University of Port Bizabeth: M.Se. thesis, 1986. p.194.

ROMER, G.S. Surf zone fish community and species response to wave energy gradient. **J. Fish Biol.**, v.36, p. 279-287, 1990.

RORZG, L.R.; RESGALLA JR, C.; PEZZUTO, P.R.; ALVES, E.S.; MORELLI, F. Ecological analysis of an accumulation process of the diatom *Anaulus* sp. in the surf zone of Navegantes beach (Santa Catarina, Brazil). **Oecologia Brasiliensis**, n.3, p.29-43, 1997.

ROSSOUW, G.J. The importance of non-teleost fishes in the surf zone with special reference to *Rhinobatos annulatus*. In: McLACHLAN. A.; ERASMUS, T. **Sandy Beaches as Ecosystems**. (1th ed.). The Hague: Dr. Junk Publishers, 1983. p.749749.

RUPLE, D.L. Occurrence of larval fishes in the surf zone of a Northern Gulf of Mexico Barrier Island. **Estuarine, Coastal and Sherf Science**, v.18, p.191-208, 1984.

SAN VICENTE, C.; SORBE, J.C. Biologie du mysidacé suprabenthique Schistomysis parkeri Norman, 1892 dans la zone sud gu Golfe de Gascogne (Plage d'Hendaye). **Crustaceana**, v.65, p.222-252, 1993.

SANT'ANNA NETO, J.L. Dinàmica atmosfèrica e o caràtertero transicional do clima na zona costeira paulista. **Revista do Departamento de Geografia Universidade de**

Sao Paulo, Sao Paulo, v.8, p.34-49, 1994.

SANT'ANNA, E.M.E.; BJONBERG, T.K.S. Seasonal dynamics of mesozooplankton in Brazilian coastal waters. **Hydrobiologia**, v.563, p.253-268, 2006.

SÂO PAULO (State). Department of the Environment. Xixovà-Japui State Park - Management Plan - Phase 1 - Consolidation of data and preliminary guidelines. Sao Paulo: Coordination of Technical Information, Documentation and Environmental Research; Forestry Institute; Botanical Institute; CEPEL-UNESP, 1997. p.74 (Environmental Documents Series).

SÂO PAULO (State). Department of the Environment. Xixovà-Japui State Park - Management Plan - Main Volume. Sao Paulo: Coordination of Technical Information, Documentation and Environmental Research; Forestry Institute; Botanical Institute; CEPEL-UNESP, 2010. p.544 (Environmental Documents Series).

SATYAMURTY, P.; NOBRE, C.A.; SILVA DIAS, P.L. Meteorology of the tropics: South America. In: KAROLY, D.J.; VICENT, D.G. (Ed.). **Meteorological Monographs**. 1998. v.27, n.49, p.119-139.

SHANKS, A.L.; MORGAN, S.G.; MACMAHAN, J.; RENIERS, A.H.M. Surf zone physical and morphological regime as determinants of temporal and spatial variation in larval recruitment. **Journal of Experimental Marine Biology and Ecology**, v.392, p.140-150, 2010.

SHANNON, C.E. A mathematical theory of communication. **The Bell System Tech J.**, v.27, p.379-423, 1948.

SHORT, A.D. Sediments and structures in beach-nearshore environments, south east Africa. In McLACHLAN, A.; ERASMUS, T. **Sandy beaches as ecosystems**. (1th ed.) The Hague: Dr. Junk Publishers, 1983.p.145-155.

SHORT, A.D. (Ed.). **Handbook of beach and shoreface morphodynamics**, Chichester: Wiley publishing chichester, 1999.

SIEBURTH, J. McN.; SMETACEK, V.; LENZ, J. (1978). Pelagic ecosystem structure: heterotrophic compartments of the plankton and their relationship to plankton size fractions. **Limnology and Oceanography**, v.23, n.6, p.1256-1263, 1978

SILVA, M.; ARAÙJO, F.G.; COSTA DE AZEVEDO, M.C.; DE SOUSA SANTOS, J.N. The nursery function of sandy beaches in a Brazilian tropical bay for group anchovies (Teleostei: Engraulidae): diel, seasonal and spatial pattern. **Journal of the Marine Biological Association of the United Kingdom**, v.84, p.229-1232, 2004.

SMITH, D.L.; JOHNSON, K.B. (2 th ed.). **A guide to marine coastal plankton and marine invertebrate larvae**. Iowa: Kendall/Hunt Pulishing Co., 1996. p. 221.

SOUZA, C.R. de G.; SUGUIO, K. Coastal erosion and beach morphodynamics along the state of Sao Paulo (SE Brazil). **Anais da Academia Brasileira de Ciências**, Rio de Janeiro, v.68, p. 405-424, 1996.

SOUZA, C.R. de G. **Coastal Drift Cells and Erosion on the Beaches of the State of São Paulo.** 1997. v.1/2. 204/186f. Thesis (Doctorate in Science, Area of Sedimentary Geology) - Institute of Geosciences, University of Sao Paulo, Sao Paulo. 1997.

SOUZA, C.R de G. Coastal erosion risk assessment, shoreline retreat rates and causes of coastal erosion along the state of Sao Paulo coast, Brazil. **Revista Pesquisas em Geociências**, Porto Alegre, v.28, n. 2, p.459-474, 2001.

SOUZA, C.R. de G.; SUGUIO, K. The coastal erosion risk zoning and the Sao Paulo state plan for coastal management. **Journal of Coastal Research**, v.35, p.530-592, 2003.

SOUZA, C.R. de G. Sandy ocean beaches of the State of Sao Paulo (Brazil): synthesis of knowledge on morphodynamics, sedimentology, coastal transport and coastal erosion. **Revista do Departamento de Geografia - USP**, Volume Especial 30 Anos. p. 307-371, 2012.

SOUZA, T.A.; CUNHA, C.M.L. The importance of drainage enrichment for morphometric analysis in coastal plain areas - the case of the Municipality of Praia Grande (SP). In: SEMINÀRIO DE PÓS GRADUAÇÃO EM GEOGRAFIA DA UNESP RIO CLARO, 2009. **Proceedings...** Rio Claro: Universidade Estadual Paulista. p.1739-1753.

SOUZA, T.A.; CUNHA, C.M.L.A.Anâlise dos atributos fisico-ambientais do Municipio de Praia Grande. **Sp. Soc. & Nat.**, Uberlândia, ano 24, n. 2, p.303-318, 2012.

STRAUB, L.M.; SALVADOR, B.; BERSANO, J.G. Temporal variation of biodiversity, abundance and secondary production of zooplankton in Pontal do Sul, PR. In: BRAZILIAN CONGRESS OF OCEANOGRAPHY, 2014, Itajai- SC. **Annals of Abstracts**. Brazilian Association of Oceanography, 2014. p. 11021103.

STUCK, K.C.; PERRY, H.M.; HEARD, R.W. Records and range extensions of Mysidacea from coastal and shelf waters of the eastern Gulf of Mexico. **Gulf Research Reports**, n. 6, p. 239-248, 1979.

STULL, K.J.; CAHOON, L.B.; LANKFORD, T.E. Zooplankton Abundance in the Surf Zones of Nourished and Unnourished Beaches in Southeastern North Carolina, U.S.A. **Journal of Coastal Research**, v.32, n.1, p.70-77, 2016.

SUGUIO, K;; PETRI, S. Stratigraphy of the Iguape-Cananéia lagoonal region sedimentary deposits, Sao Paulo, Brazil. Part I: Field observations and grain size analysis. **Bol. IG-USP**, v.4, p.1-20, 1973.

SUGUIO, K; MARTIN, L. **Geological map: Folha Santos**. Sao Paulo: DAEE/USP/FAPESP, 1978. Scale 1:100.000.

SUGUIO, K. (Ed.). **Dictionary of Marine Geology: with corresponding terms in English, French and Spanish.** Sao Paulo: T.A. Queiroz, 1992.

SUTHERS, I.M.; RISSIK, D. (Ed.). **Plankton: A guide to their ecology and monitoring for water quality.** Australia: Csiro Publishing, 2009.

TAKAHASHI, K.; KAWAGUCHI, K. Diel and tidal migrations of the sandburrowing mysids, Archaeom ysis kokuboi, A. japonica and Iiella ohshimai, in Otsuchi Bay, northeastern Japan, **Marine Ecology Progress Series**, v.148, p.95107, 1997.

TAKAHASHI, K.; KAWAGUCHI, D. Diet and feeding rhythm of the sandburrowing mysids *Archaeomysis kokuboi* and *A. japonica* in Otsuchi Bay, northeastern Japan. **Marine Ecology Progress Series,** v.162, p.191-199, 1998.

TAKAHASHI, K.; HIROSE, T.; KAWAGUCHI, K. The importance of intertidal sand-burrowing peracarid crustaceans as prey for fish in the surf-zone of a sandy beach in Otsuchi Bay, northeastern Japan. **Fish Sci**, v.65, p.856-864, 1999.

TESSLER, M. G. **Quaternary sedimentary dynamics on the south coast of São Paulo**. 1988. 276f. Thesis (Doctorate in Sciences, Area of Sedimentary Geology) - Institute of Geosciences, University of Sao Paulo, Sao Paulo. 1988.

TESSLER, M.G.; MAHIQUES, M.M. Oceanic processes and seabed physiography. In: TEIXEIRA, W.; TOLEDO, M.C.M.; FAIRCHILD, T.R.; TAIOLI, F. (Org.). **Deciphering the Earth**. (1 Ed.). Sao Paulo: Oficina de Textos, 2000. p.264286.

TESSLER, M.G.; GOYA, S.C.; YOSHIKAWA, P.S.; HURTADO, S.N. Erosao e progradaçao do litoral do Estado de Sao Paulo. In: MUEHE, D. (Org.) **Erosao e progradaçâo do litoral brasileiro**. (1 Ed.) Brasilia: Ministérios do Meio Ambiente, 2006.

TOMMASI, L.R. 1979. **Ecological considerations on the estuarine system of Santos, Sâo Paulo**. 1979. v.2. Dissertation (Livre-Docência) - Oceanographic Institute, University of Sao Paulo, Sao Paulo. 1979.

VASCONCELLOS, R.M.; SANTOS, J.N.S.; SILVA, M.A.; ARAÙJO, F.G. Effect of the degree of exposure to waves on the juvenile fish community in sandy beaches of the Municipality of Rio de Janeiro, Brazil. **Biota Neotrop**. v.7, n.1, p.93100, 2007.

VAZZoLER, A.E.M. **Biologia da reproduçâo de peixes Teleósteos: teoria e pràtica**. Maringà: EDUEM, 1996.

VEGA-PÉREZ, L.A, Estudo do zooplâncton da regiao de Ubatuba, Estado de Sao Paulo. **Special publication of the Oceanographic Institute of the University of São Paulo**, v.10, p.65-84, 1993.

VILANO, W.F. **Biogeographic and Paleogeographic Aspects of Ubatuba Rocky Cliffs - Sao Paulo**. 2013. 160f. Dissertation (Master of Science) - Institute of Physical Geography of the Faculty of Philosophy, Letters and Human Sciences, University of Sao Paulo, Sao Paulo. 2013.

WAINER, I.; TASCHETTO, A.S. Climatology in the region between Cabo de Sao Tompe (RJ) and Chui (RS), diagnosis for periods related to the fishing surveys of the REVIZEE program. In: ROSSI-WONGTSCHOWSHI, C.L.D.B.; MADUREIRA, L.A.P. **O ambiente oceanogràfico da plataforma continental e do talude na Regiào Sudeste-Sul do Brasil.** Sao Paulo: EDUSP. 2006. p. 121160.

WARMAN, C.G.; O'HARE, T.J.; NAYJOR, E. Vertical swimming in wave-induced currents as a control mechanism of intertidal migration by sand-beach isoped. **Marine Biology,** v.111, p.49-54, 1991.

WATKIN, E.E. Observations on the night tidal migrant Crustacea of Kames Bay. **Journal of the Marine Biological Association of the United Kingdom**, v.25, p.81-96, 1941.

WATT-PRINGLE, P.; STRYDOM, N.A. Habitat use by larval fishes in a temperate South African surf zone. **Estuar. coast. Shelf Sci**., v.58, p.765-774, 2003.

WEBB, P.; WOOLDRIDGE, T.H. *Feeding of Mesopodopsis slabben* (*Crustacea, Mysidacea) on naturally occurring phytoplankton*. **Marine Ecology Progress,** v.38, p.15-123, 1987.

WEBB, P.; WOOLDRIDGE, T.H. Diel horizontal migration of *Mesopodopsis slabberi* (Crustacea: Mysidacea) in Algoa Bay, southern Africa. **Marine Ecology Progress Series**, v.62, p.73-77, 1990.

WHITFIELD, J.L.B. Ichthyoplankton in a Southern African Surf Zone: Nursery Area for the Postlarvae of Estuarine Associated Fish Species? **Estuarine, Coastal and Shelf Science,** v.29, p.533-547, 1989.

WOOLDRIDGE, T.H. Zonation and distribution of the beach mysid, Gastrosaccus psammodytes (Crustacea: Mysidacea). **Journal of Zoology**, London, v.193, p.183-189, 1981.

WOOLDRIDGE, T.H. Ecology of beach and surf zone mysid shrimps in the eastern Cape, South Africa. In: McLACHLAN. A.; ERASMUS, T. **Sandy Beaches as Ecosystems**. (1th ed.). The Hague: Dr. Junk Publishers, 1983. p.449-460.

WOOLDRIDGE, T.H. The spatial and temporal distribution of mysid shrimps and phytoplankton accumulations in a high energy surfzone. **Vie et Milieu Life and Environment**, v.39, p.127-33., 1989.

WRIGHT, L.D.; GUZA, R.T.; SHORT, A.D. Dynamics of a high energy dissipative surf zone. **Marine Geology**, v.45, n.1-2, p.41-62, 1982.

WRIGHT, L.D.; SHORT, A.D. Morphodynamic the beaches and surf zones in Australia. In: KOMAR, P.D. (Ed.). **Handbook of coastal processes and erosion.** Florida, USA: CRC Press, 1983. p.35-64.

WRIGHT, L.D.; SHORT, A.D. Morphodynamic variability of surf zones and beaches: a synthesis. **Marine Geology**, v.56, p.93-118, 1984.

YU, O.H.; SUH, H.L.; SHIRAYAMA, Y. Feeding ecology of three amphipod species *Synchelidium lenorostralum*, *S. trioostegitum* and *Gitanopsis japonica* in the surf zone of a sandy shore. **Marine Ecology Progress Series**, v.258, p.189-199, 2003.

Annex

Annex 1: Studies on the zooplankton of the upwelling zone carried out along the Brazilian coast (* Conference abstracts; ** Monograph-Graduation; *** Dissertation-Master's degree; **** Thesis-Doctorate; Δ Published articles). Source: Modified from IBGE cartographic base.

Autor		Fauna Estudada	Ênfase	Objetivo	Local	Período de Amostragem
Bersano, 1994	***	Fauna zooplanctônica	-	Análise da composição e variação temporal do zooplâncton	Praia do Cassino (RS)	Primavera/ 1990 e Verão/1991
Bersano *et al.*, 1995	*	Crustacea: Mysidacea	-	Análise da composição e variação temporal de misidáceos	Praia do Cassino (RS)	Primavera/ 1990 e Verão/1991
Bersano et al., 1997		Fauna zooplanctônica	-	-	Praia do Cassino (RS)	-
Gorri, 2006	****	Crustacea: Copepoda	-	Análise da composição e variação temporal de copépodos	Praia do Cassino (RS)	-
Borzone *et. al.*, 2007	Δ	Crustacea: Mysidacea	-	Análise temporal da composição, abundância e distribuição de misidáceos no substrato	Praia de Pontal do Sul (PR)	Mensal (Março/2001)
Calil & Borzone, 2008	Δ	Crustacea: Mysidacea	*Metamysidopsis neritica*	Análise das características populacionais e biologia reprodutiva da espécie	Praia de Pontal do Sul (PR)	Mensal (Agosto/1999 a Julho/2000)
Avila *et al.*, 2009	Δ	Fauna zooplanctônica	Crustacea: Copepoda	Análise temporal da composição, abundância, diversidade e biomassa da fauna zooplanctônica	Praia de Tramandaí (RS)	Quinzenal (Agosto/ 2005 a Agosto/ 2006)
Felipe & Bersano, 2010	*	Crustacea: Mysidacea	-	Análise nictimeral da composição, abundância e distribuição de misidáceos no plâncton e no substrato	Praia de Pontal do Sul (PR)	Mensal (Março e Outubro/2010)
Felipe, 2010	**					
Resgalla, 2010	Δ	Chaetognatha	*Sagitta friderici*	Reconhecimento de parâmetros polupacionais da espécie e comparação com populações de outras regiões	Praia do Cassino (RS)	Quinzenl (Outubro/1990 a Abril/1992)
Costa *et al.*, 2011	Δ	Fauna zooplanctônica	-	Análise da composição e variação sazonal do zooplâncton e sua relação variáveis abióticas	Praia de Ajuruteua/PA	Bimensal (Fevereiro a Dezembro/2008)
Grassi *et al.*, 2011	*	Crustacea: Copepoda	-	Análise da variação temporal de copépodos	Praia de Pontal do Sul (PR)	-
Pinheiro *et al.*, 2011	Δ	Crustacea: Copepoda	-	Análise da composição e variação temporal de copépodos	Praia de Ajuruteua/PA	Mensal (Agosto/2004 a Julho/2005)
Costa *et al.*, 2014a	*	Fauna zooplanctônica	-	Análise da composição e variação sazonal do zooplâncton e sua relação variáveis abióticas	Praia de Ajuruteua/PA	Bimensal (Dezembro/2011 a Outubro/2012)
Costa *et al.*, 2014b	*	Crustacea: Copepoda	-	Análise da composição e variação sazonal de copépodos e sua relação variáveis abióticas	Praia de Ajuruteua/PA	Bimensal (Dezembro/2011 a Outubro/2012)
Straub *et al.* 2014	*	Fauna zooplanctônica	Crustacea: Copepoda	Avaliar as variações temporais de diversidade e abundância do zooplâncton	Praia de Pontal do Sul (PR)	Quinzenal (Março/ 2012 a Março/ 2013)

Annex 2: Sensal abundance of the taxa caught over two sampling years. **Est.** = Life stage; Ad. = Adult; Juv. = Juvenile; La. = Larval; Pl. = Post-larva; Meg. = Megalopa; Nau. = Nauplio; **Cla. =** Classification according to capture frequency: R = Resident; M = Migratory; E = Sporadic; M/E= Migratory or Sporadic; **Abundance:** *: 0.1 - 5 org. иг^3; **: 5.1-20 org. иг^3; ***: 20.1 - 100 org. m\ ****: 100.1 - 500 org. m\ *****(:) 500.1 - 1,000 org. m\ Δ: 1,000.1 - 5,000 org. m\ ΔΔ: 5,000.1 - 10,000 org. m\ ΔΔΔ: 10,000.1 - 30,000 org. m\ ΔΔΔΔ: 30,000.1 - 60,000 org. m (3).

Táxon	Est.	1º Ano												2º Ano												Clas.
		Jun'13		Ago'13		Out'13		Dez'13		Fev'14		Abr'14		Jun'14		Ago'14		Out'14		Dez'14		Fev'15		Abr'15		
		D	N	D	N	D	N	D	N	D	N	D	N	D	N	D	N	D	N	D	N	D	N	D	N	
Tintinnina							**		*												*					M/E
Foraminifera																										
Globigerinidae																										
Globigerinoides ruber	Ad					*	*	*	*	*	***			**	*	*		*	*	*	*	**	*	***	*	R
Globigerina bulloides	Ad								*	*	*															E
Orbulina universa	Ad																									E
Foraminifera N.I.	Ad									*	**			*	*	*										R
Cnidaria																										
Hydrozoa																										
Tubulariidae	Juv	**			*																					E
Campanulariidae																										
Obelia sp.	Juv				*	*		*	*	*		*		*		*					*			*	*	R
Abylidae	Ad				*																					E
Apolemiadae	Ad							*																		E
Diphyidae	Ad							**	*				*					*		*	*	*	*			M/E
Physalidae																										
Physalia sp.	Juv				*																					E
Rhizophysidae																										
Rhizophysa sp.	Juv				*																					E
Geryoniidae																										
Liriope tetraphylla	Ad			*						**																E
Hidrozoa N.I.	Juv	**	***	*	**	*	**	*	*	**	*	*	*	*	*	*		*	*	*	*			*	***	R
Cubozoa	Ad		*														*			*	*			*		E
Scyphozoa	Ad				*																					E
Ctenophora	Juv													*	*	*	*							**	*	E
Bryozoa	La		***	*	*	*	*	*	*		*	*				*		*						*	**	R
Mollusca																										
Gastropoda	La	***	***	**	**	***	****	***	***	**	***	***	**	*****	***	**	***	**	**	**	***	****	**	****	*****	R
Bivalvia	La	***	***	*	***	**	**	**	**	***	***	**	*	****	*	*	**	*	***	**	**	***	***	****	****	R
Polychaeta																										
Chrysopetalidae	La					*	**			*				*				*								E
Glyceridae	La																				*					E
Nereididae	La						*	*		*	*	*						*								M/E
Phyllodocidae	La					*	*																			E
Syllidae	La						*																*			E
Owenidae	La					*																				E
Spionidae	La	*	***	*	*	**	***	***	*	***	**	**	**	****	*	*	*	*	*	*	*	*	*	*****	***	R
Magelonidae	La					*	*																			E
Pectinariidae	La						*		*																	E
Terebellidae	La									*			*											*		M/E
Polychaeta N.I.	La						*																			E

Continued Annex 2: Monthly abundance of taxons caught over two years of sampling. **Est.=** Life stage; Ad. = Adult; Juv. = Juvenile; La. = Larval; Pl. = Post-larva; Meg. = Megalopa; Nau. = Nauplio; **Cla. =** Classification according to catch frequency: R = Resident; M = Migratory; E = Sporadic; M/E= Migratory or Sporadic; **Abundance:** *: 0.1 - 5 org. m \ **: 5.1-20 org. m \ ***: 20.1 - IOO org. m \ ****: 100.1 - 500 org. m ^{3}: *****: 500.1 - 1,000 org. m \ Δ: 1,000.1 - 5,000 org. m \ ΔΔ: 5,000.1 - 10,000 org. m \ ΔΔΔ: 10,000.1 - 30,000 org. m \ ΔΔΔΔ: 30,000.1 - 60,000 org. m^3.

Táxon	Est.	1º Ano												2º Ano												Clas.
		Jun'13		Ago'13		Out'13		Dez'13		Fev'14		Abr'14		Jun'14		Ago'14		Out'14		Dez'14		Fev'15		Abr'15		
		D	N	D	N	D	N	D	N	D	N	D	N	D	N	D	N	D	N	D	N	D	N	D	N	
Copepoda																										
Calanoida	Ad	Δ	Δ	*****	*****	****	****	****	*****	Δ	ΔΔ	ΔΔ	Δ	Δ	****	****	Δ	****	****	Δ	Δ	Δ	Δ	ΔΔ	ΔΔΔΔ	R
Cyclopoida	Ad	***	****	****	****	***	***	**	***	****	****	***	**	***	***	**	***	**	**	***	****	****	**	****	****	R
Harpacticoida	Ad	***	***	***	**	*	***	*	*	***	****	*	*	*	*	**	**	*	*	*	*	***	***	**	***	R
Poecilostomatoida	Ad	*	*	**	**	**	***	*	*	***	***	****	****	*	*	**	***	***	*	**	*	****	**	*****	*****	R
Copepoda	Nau	***	**	**	*	**	**	**	*	*	**	*	*	*		*	**	*	*	*	*	*	*	**	**	R
	Nau						*	**	**	***	*	*	**	**			**	*	*	*	*	*	*	**	****	R
Cirripedia	La	*	*	*	*	*	*	***	***	***	****	*	*	**			*	*		***			*	***	***	R
Cladocera																										
Podonidae																										
Evadne sp.	Ad					*	*																			E
Pseudevadne sp.	Ad	*				**	***			*	**							*			*			****	*****	M/E
Podonidae N.I.	Ad		***			**	***			*		*	*						*						**	R
Sididae																										
Penilia sp.	Ad		****	*	*	*	*	*	**	***	**	*	*					*	*	***	**			Δ	Δ	R
Amphipoda																										
Gammaridea	Ad	*		*	*	*	***	*	**	*	*	*	*	*	*	*	*	*	*		*	*	*	***	**	R
Hyperidea																										
Hyperoche sp	Ad				*												*									M
Isopoda																										
Bopyridae	Ad	***	**	*	*			*	*	*		*	*	*		*							*		*	R
Idoteidae																										
Edotia sp.	Ad							*	*						*											E
Isopoda N.I.	Ad		**				*								*						*	**	**	**	***	R
Mysida																										
Bowmaniella brasiliensis	Juv/Ad				*			**	**		*		*	*	*		**		*		**					R
Metamysidopsis elongata	Juv/Ad	**	****	***	*****		*	*	****		*	***	***	***	*		***	*	*	*	**	***	*	**	****	R
Decapoda	Zoea	*	*	*	*	*	*	*	*	*		*	*	*	*	*	*	*	*	*	*	*		*	*	R
Brachyura	Meg		*		*		*		**		*	*	**	*	*	*	**		*		*		**	*	**	R
Caridea	Zoea		*	*	*		*	*						*				*		*	*	***	*			R
Penaeidae	Pl		*	*	*					*	*	*	*	*	*	*	*	*	*		*		*		**	R
Thalassinidae	Zoea		*						***					*					*		***	*	**			R
Luciferidae																										
Lucifer sp.	Ad														*						*					E
Euphausiacea	La		*			*	*	***	*	*	*	*		*				*	*			*		*	*	R
Echinodermata																										
Asteroidea	La				*	*							*			*		*	*							M
	La	**	**			***	***	*	*			***	***			*	*	*	*				*	Δ	Δ	R
Echinoidea	Juv	*	**			*	**			*								*						**	***	E
Holothuroidea	La		**				*	*					*					*								E
Ovo Invertebrado N.I.		**	**	*	*	**	***	***	**	***	*	*	*	*		*	*	*		***	*	**				-

Continued Annex 2: Monthly abundance of taxons caught over two years of sampling. **Est.=** Life stage; Ad. = Adult; Juv. = Juvenile; La. = Larval; Pl. = Post-larva; Meg. = Megalopa; Nau. = Nauplio; **Cla. =** Classification according to capture frequency: R = Resident; M = Migratory; E = Sporadic; M/E= Migratory or Sporadic; **Abundance:** *: 0.1 - 5 org. m \ **: 5.1 - 20 org. m \ ***: 20.1 - 100 org. m \ ****: 100.1 - 500 org. m ^{3}: *****: 500,1 - 1.000 org. m \ Δ: 1.000,1 - 5.000 org. m \ ΔΔ: 5.000,1 - 10.000 org. m \ ΔΔΔ: 10.000,1 - 30.000 org. m \ ΔΔΔΔ: 30.000,1 - 60.000 org m^3.

Táxon	Est.	1º Ano												2º Ano												Clas.
		Jun'13		Ago'13		Out'13		Dez'13		Fev'14		Abr'14		Jun'14		Ago'14		Out'14		Dez'14		Fev'15		Abr'15		
		D	N	D	N	D	N	D	N	D	N	D	N	D	N	D	N	D	N	D	N	D	N	D	N	
Chaetognatha																										
Sagittidae																										
Sagitta sp.	Ad	**	**	**	**	**	*	*	*	**	**	*		**	*	**	*	*	**	*	***	***	*	*	**	R
Chordata Tunicata																										
Oikopleuridae																										
Oikopleura sp.	Ad	***	**	*		****	****	*	***	*****	****	*	**		*	*	*	***	**	***	**	*	**	****	***	R
Ascidiacea	La	***	*				*		*																	E
Salpidae																										
Thalia sp.	Ad	**	**	*	*	*			**				**			*		*		*	*				*	M/E
Doliolida	Ad		*							**	*							*								E
Chordata Vertebrata																										
	Ovo	*			*	*	***	*	**	*	***	*	*			*	*	*	**	**	**	**	**	**	*	-
Actinopteri	La		*			*	*	*	*	*	*		*		*			*		*				*	*	R
Atherinella blackburni	Juv																						*		*	E
Menticirrhus littoralis	Juv						*		*				*							*						M/E
Oligoplites sp.	Juv						*										*						*			M/E
Conodon nobilis	Juv						*			*							*		*							M/E
Mugil brevirostris	Juv																*									E
Mugil curema	Juv								*		*		*													M/E
Symphurus sp.	Juv								*								*									E
Linguado sp.1	Juv								*			*	*												*	M/E
Linguado sp.3	Juv																								*	E
Linguado sp.4	Juv																								*	E
Anchoviella lepidentostole	Juv								*	*			*												**	M/E
Exocoetidae	Juv																			*	*					E
Gobiidae	Juv																				*					E

More Books!

I want morebooks!

Buy your books fast and straightforward online - at one of world's fastest growing online book stores! Environmentally sound due to Print-on-Demand technologies.

Buy your books online at
www.morebooks.shop

Kaufen Sie Ihre Bücher schnell und unkompliziert online – auf einer der am schnellsten wachsenden Buchhandelsplattformen weltweit! Dank Print-On-Demand umwelt- und ressourcenschonend produzi ert.

Bücher schneller online kaufen
www.morebooks.shop

info@omniscriptum.com
www.omniscriptum.com

OMNIScriptum

MIX
Papier aus verantwortungsvollen Quellen
Paper from responsible sources
FSC® C105338

FSC
www.fsc.org

Printed by Books on Demand GmbH, Norderstedt / Germany